PRACTICAL

PLANE GEOMETRY.

BY

E. S. BURCHETT,

LECTURER ON GEOMETRICAL AND PERSPECTIVE DRAWING AT THE NATIONAL ART
TRAINING SCHOOL, SOUTH KENSINGTON; AT HAILEYBURY COLLEGE;
AT ST. MARK'S COLLEGE, CHELSEA; ETC.

LONDON AND GLASGOW:
WILLIAM COLLINS, SONS, AND COMPANY.
1876.

PRACTICAL

PLANE GEOMETRY.

INTRODUCTION.

THIS work has been compiled on account of a strongly-felt want of a more complete text book upon this subject, the author having been constantly asked, for some years past, by his own and other students, why he did not write a book which would be on the same level as the more advanced lecturing and teaching of the present time.

The study of Geometry has made such rapid strides in general appreciation during the last twenty years, that those books which were previously quite sufficient, have become wholly inadequate to the demands made upon them, and several works have been published, which have considerably developed the subject, some one section, and some another, but so far as the author's knowledge extended, there seemed still a very large field of usefulness for a work which would unite in itself, as far as possible, the matter to be found in a number, and which would more completely develop some of the sections, while it at the same time so arranged and methodized the whole subject, as to bring it more easily within the grasp of the student.

The author has pleasure in acknowledging his indebtedness to three works on this subject in particular:

1st, *Geometry: Plane, Solid, and Spherical,* published in 1830, under the superintendence of the Society for the Diffusion of Useful Knowledge.

2nd, To another work published later by the same Society, by Thomas Bradley; and

3rd, To a book on *Geometry and Plan Drawing,* by C. W. Pasley, Major-General, dated 1843; and to Alexander Macdonald, Esq., of the Oxford School of Art, he is indebted for the solution of Problem 255.

It should be understood, that while the great majority of the problems contained in this book, are worked by methods which are capable of Euclidic demonstration, that there are some few which are not so, particularly among those relating to the construction of polygons, the last three problems on the contact of circles, and some relating to the areas of circles; in these cases methods have been chosen which are quite practically correct, it having been felt, that in a work which only professed to be practical, it would be better to have a method which was theoretically inadmissible, than one which was practically inaccurate.

Many students, in glancing over this book, may be impressed with the idea, that there is so much more than they want for preparation for an Elementary Examination, or for the purposes of practical art, that it would be impossible for them to select what would be most useful or necessary to study; but this difficulty will be to a great extent obviated if the student, taking the part on Plane Geometry, confines his attention to the earlier plates of each section, rather than completing each section as he advances; which is perhaps an amount of study, requisite only for the more advanced examinations, or a more complete mastery of the subject.

Following out this last thought, a teacher might recommend a student who was preparing for what is generally understood as a "Second Grade Examination," to pass over Plates 7, 8, 12, 17, 18, 19, 23, 24, 25, 26, 27, 28, 35, 36, and the latter half of the section on areas, reserving these for a future study; but it should be impressed upon the student's mind, that the part on projection has been appended to this work especially to meet the requirements of the more extended range of the Second Grade Examination of the present day.

E. S. B.

20 BROMPTON SQUARE,
October, 1875.

NOTES TO PLATE I.

Note 1.—To avoid the frequent rewriting of some long words, the following abbreviations, which will be easily remembered, will in future be used in this work.

For Parallelogram—Pgram.
" Parallel,   —Parl.
" Perpendicular—Perpr.
" Proportional, —Propnl.
" Equilateral,   —Eqtrl.
" Horizontal,   —Hortl.
" Vertical,   —Vertl.

Note 2.—Degrees are denoted by a small circle after the number, thus—360° or 90°.

Note 3.—The student who has had no previous study of Euclid should observe the following points, that may be culled from the First Book :—

1st.   The three angles of a triangle are, together, equal to two right angles, invariably.

2nd.   The triangle has the same number of equal sides that it has equal angles.

3rd.   The equal angles are always opposite the equal sides.

4th.   If the side of a triangle is produced, the exterior angle so produced is equal to the two interior, and opposite angles of the triangle.

5th.   The greater sides are always opposite the greater angles.

6th.   Any two sides of a triangle must be together greater than the third side.

7th.   A triangle can only have one obtused or one right angle in it.

CONTENTS.

DETAILED CONTENTS OF PART II.

FIRST SECTION.

Plate IV.

Plate V.

Plate VI.

Plate VII.

PLATE VIII.

SECOND SECTION.—CONSTRUCTIONS.

PLATE IX.

PLATE X.

PLATE XI.

PLATE XII.

PLATE XVII.

PLATE XVIII.

PLATE XIX.

THIRD SECTION.—CURVES.

PLATE XX.

PLATE XXI.

PLATE XXII.

FOURTH SECTION.—CIRCLES.

FIFTH SECTION.—FIGURES IN RELATION TO FIGURES.

SIXTH SECTION.—PROBLEMS RELATING TO AREAS.

PLANE GEOMETRY IS THE SCIENCE WHICH DEALS WITH THE RELATIONS OF LINES AND SHAPES UPON A FLAT SURFACE OR PLANE.

DESCRIPTIVE DEFINITIONS.

A Point is a position in space, which may be indicated on a plane, by a simple mark or dot (1), or the intersection of two lines (2), or the extremity of a line.

A Line is an indication of division, or boundary, generated by the motion of a point, and has only the property of length.

There are straight or right lines (3), curved lines (4), and compound lines, composed of right and curved lines. Compound lines may be continuous (5), when the curve continues the right line: or broken (6), when the right and curved parts of the line, form an angle.

Lines are represented in Geometrical drawing, as a strong line (7), a thin, or fine line (8), a dotted line (9), and a chain-dotted line (10); the two first are called solid lines.

An Angle. When two right lines meet in a point, they form an angle (11), which is an indication of a portion of the space round the point; this space is, for the purpose of measurement of angles, supposed to be divided into 360 equal parts, or angles, which are called degrees. The lengths of the lines forming the sides of the angle have nothing to do with its measurement. When two lines cross in such a way as to form four equal angles (12), they are said to be at a right angle, or perpendicular to each other; an angle of one fourth of the space, or 90 degrees, being always called a *Right Angle.* An angle that contains less than a right angle is called acute (13), and an angle that is greater than a right angle is called obtuse (13A).

Lines that are at an unvarying distance from each other, and thus would never meet if produced, or continued in either direction, are said to be *parallel* to each other.

A Line, or combination of lines enclosing space, constitutes a *figure. Figures* are either rectilinear,—composed of right lines, or curvilinear,—composed of curved lines. Right-lined figures are divided into *Triangles,* or *Trilaterals,* having three sides; *Quadrangles,* or *Quadrilaterals,* having four sides; and *Polygons,* or *Multilaterals,* having more than four sides.

Of triangles, the *Equilateral* (14), has three sides and three angles equal; the *Isosceles* (15), has two sides and two angles equal; and the *Scalene* (16), has all its sides and angles unequal. When an isosceles triangle has one right angle it is called a *right-angled Isosceles* (17), and similarly, we may have a *right-angled Scalene* (18).

Any side of a triangle may be called its *Base,* and a line drawn perpendicular to the base, from the opposite angle, is its *Altitude* (14A); it is sometimes necessary to extend the base to show the altitude, as in (16B).

All *Quadrilaterals* having their opposite sides parallel are called *Parallelograms,* and all others are *Trapezia.*

When the angles of a parallelogram are right angles, it is a *Rectangle,* or *Right Parallelogram;* and when the angles are not right angles it is an *Oblique Parallelogram;* of rectangles, a *Square* (19), has the four sides equal, and all the angles right angles; an *Oblong* (20), has only its opposite sides equal, and all its angles right angles; of oblique parallelograms the *Rhombus,* or *Lozenge* (21), has all its sides equal, and the opposite angles equal; and the *Rhomboid* (22), has its opposite sides and opposite angles equal.

Trapezia are divisible into the *Trapezium* (23), which has no sides parallel, the *Trapezoid* (24 and 26), which has two sides parallel, and the *Trapezion* (25), which has adjacent pairs of sides equal.

Lines joining the opposite angles of quadrilaterals are called *diagonals* (22A), and in parallelograms the intersection of these lines gives the *centre* of the figure. (See Notes to Introduction.)

PLATE I.

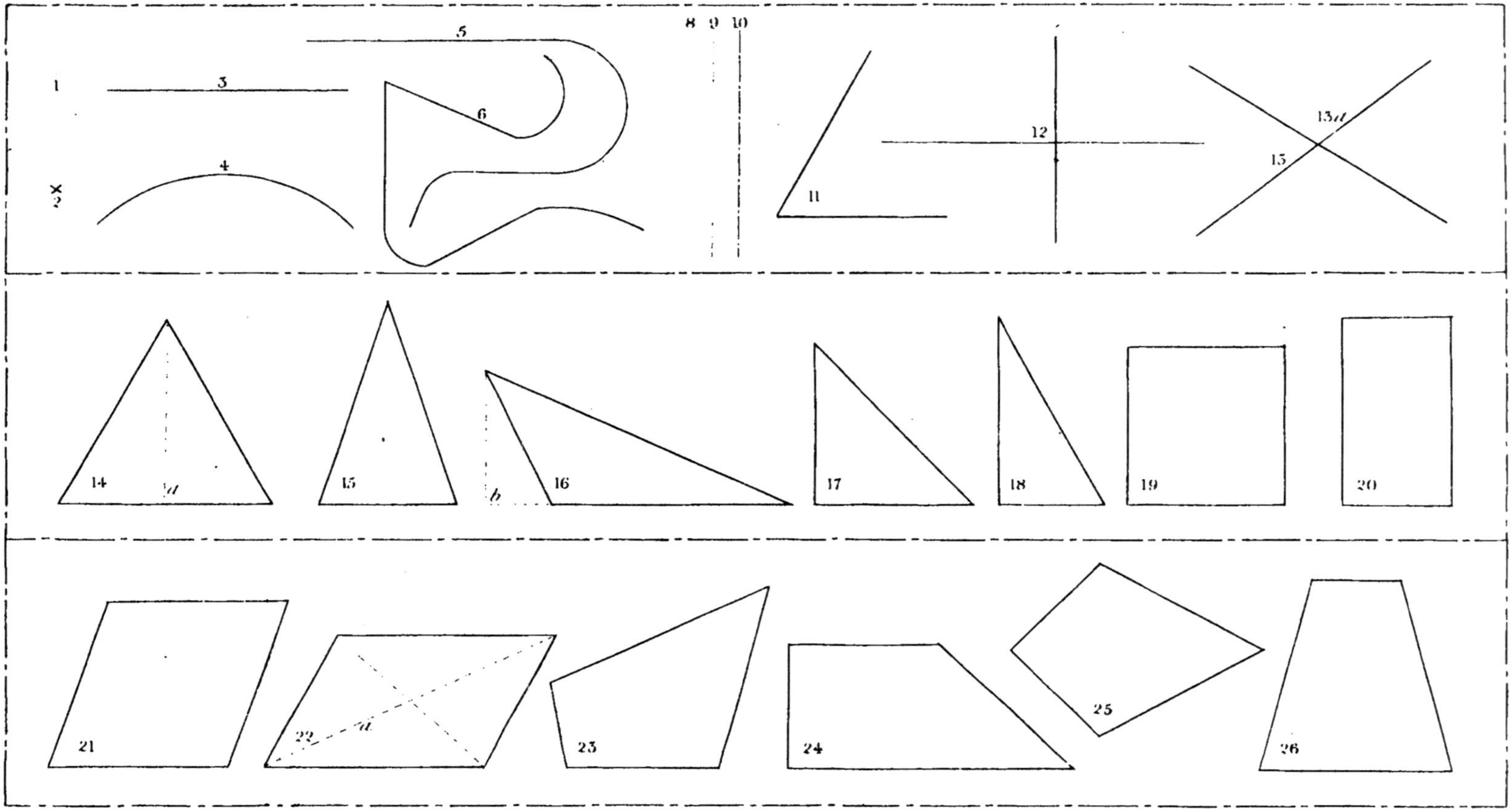

FIGURES OF MORE THAN FOUR SIDES.

Polygons, are either *regular*, having all their sides and angles equal, or *irregular*. Polygons are named according to the number of their sides or angles, as—

A Pentagon—five-sided.	A Hexagon—six-sided.
A Heptagon—seven-sided.	An Octagon—eight-sided.
A Nonagon—nine-sided.	A Decagon—ten-sided.
An Undecagon—eleven-sided.	A Dodecagon—twelve-sided.

Irregular Polygons must always be mentioned as such, or otherwise a regular polygon is understood. Figures 1, 2, 3, 4, 5, 6, 7, and 8, show regular polygons, and 9, 10, 11, and 12 show varieties of irregular polygons.

Any right line joining two angles of a polygon, and not a side, is a *diagonal*; and any right line drawn through the centre of a regular polygon, and continued to touch its sides, is a diameter. The *centre* is a point equidistant from all its sides: the distance from the centre to the side of a regular polygon is called its *apothem*.

The *Geometric curvilinear figures*, are the *Circle* (13), and the *Ellipse* (18).

The *Circle* is a figure of one line, generated by the motion of a point, moving in a plane, at an invarying distance from another point, which is the centre of the circle; from this all right lines, from the centre to the circle, are equal; they are called *radii*, or, in the singular, *radius*. Two radii in a right line form a *diameter* of the circle, and the halves as divided by the diameter are called semicircles.

Any portion of the curve of a circle is called an *Arc*, and any right line joining two points in the circle is called a *Chord* (13a). When a *chord* is drawn joining any two points in a circle, it divides it into two parts, or *Segments*; these are called *major* (14), or *minor* (16) segments, according as they contain more or less than a semicircle. Two radii drawn to the extremities of any arc constitute, with the arc, a *Sector* (17). Two lines drawn from any point in the arc to the extremities of the chord, form an angle, which is the angle of the segment; and we frequently use the expression, a segment of a certain angle, to signify a segment, to contain such a certain angle. (See Note.)

An *Ellipse* (18), is a figure of one continuous curve, generated by the motion of a point in a plane, the sum of whose distances from two fixed points is always the same; these fixed points are called the *foci*; a right line drawn through these foci, and produced to the curve at both ends, is the *major axis*, or *transverse diameter* of the ellipse; and a line through the middle of the transverse diameter, perpendicular to it, extended to the curve on either side, is the *minor axis*, or diameter *conjugate* to the transverse. The point of intersection of the axes, is the centre of the ellipse, and any right line drawn through the centre, cuts the ellipse in two equal parts, and forms a diameter.

It will be well for the student to bear in mind the following properties of the circle, which may be demonstrated by Euclid:—

1st. All the angles of the same segment are equal to each other.

2nd. The angle of a semicircle (15), is a right angle.

3rd. The angles of major segments (14), are always acute, and the angles of minor segments (16), are always obtuse.

4th. The angle formed in any segment, by lines drawn from the extremities of the chord to its centre, is always double the angle of the segment.

PLATE II.

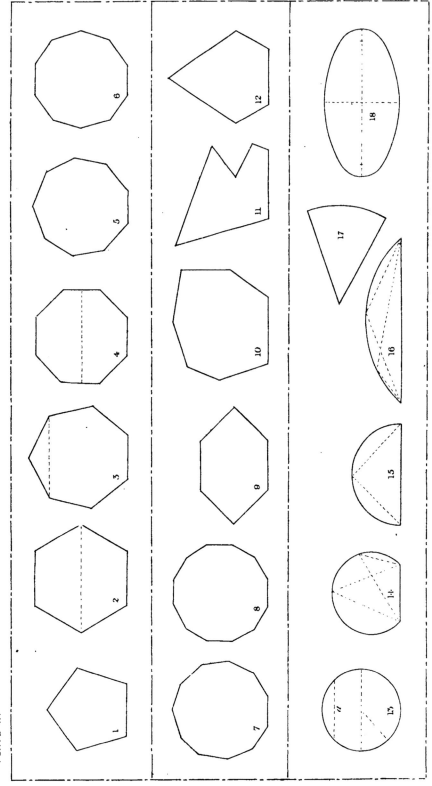

FIGS. 1 and 2 represent two varieties of the same instrument, called *pro-tractors*, one of which is found in every set of any importance; they are both used for the same purpose, i. e., to measure angles; this is accomplished by placing the base A B, upon any given line, with the centre point, C, which is always marked, at the point at which it is required to construct the angle; then, it will be observed, that the protractors have each two lines of figures, one from left to right, and one from right to left; these figures indicate the number of degrees, the degrees being marked by the lines that, though not drawn through, to economise space, are radiating from the centres C and C. If a mark is then made upon the paper, opposite any one of these figures, as mark D at 30°, and the protractor removed, and a line drawn from point C, through point D, we shall have an angle of 30°, with the line A B to the left; or, if we mark a point at the line of 25°, as E, and draw a line from C, we shall have an angle of 25°, with the line A B to the right. The form of the protractor makes no difference to the angle, since in both cases we have the lines equally radiat-ing from the centre, but cut off at different points for convenience. Fig. 1 is the most convenient form, because it gives us, at the same time, a set of plain scales, in the space that is generally lost on the semicircular protractor.

FIG. 1.—The *plain scales* are simply a number of equal parts, marked off on parallel lines, and subdivided, so as to represent feet and inches, in a certain ratio to an actual foot of measurement. Thus, scale Z gives us a number of quarters of an inch, with one divided into twelve parts. This is a scale of a quarter of an inch to a foot, or $\frac{1}{48}$ of actual measurement, though it may be also used as a quarter scale, or one-quarter of an inch to an inch. Similarly, scale Y gives us the half-inch divided in the same way, and to be used as a scale of half an inch to a foot, or $\frac{1}{24}$ of actual measurement. So also, scale X deals with three-quarters of an inch, and scale W with one inch, the last being, of course, as one inch to a foot, or $\frac{1}{12}$ actual size. It should be observed that the inches, as represented, are marked from zero to the left, and the feet from zero to the right, thus enabling us to measure off with the compasses a measurement of feet and inches, as 4′8″ from points M to N on each scale. We have also in Fig.

1, the scale marked *Chds.* or chords; this is another method for measuring angles, but it would scarcely be used by any one having a protractor; and it is curious that we rarely find a *scale of chords* given on any other instrument. By a scale of chords we mean a series of points, representing the lengths of the chords of the various arcs of a certain circle; it may be constructed in the following manner:—

FIG. 4.—A quarter of a circle, A B, is divided into the required number of degrees, as 10, 20, 30, &c., a right line, A B, is drawn; and from point A as centre a series of arcs are described, through the divisions on the quadrant, to cut the right line, as in points 10, 20, 30, &c., and this right line, with its divisions, is the scale of chords. It will be readily observed that the length of the line A B must vary, as the quadrant is greater or less; therefore, the divisions upon it will vary, and they will be only relative measurements, *i.e.*, relating only to that particular circle; the scale of chords is made available by remembering that the radius of a circle is exactly equal to the chord of one sixth of the circle, or 60°: so that if we had to measure off an angle of 25°, for example, with the scale marked on our protractor, we must describe an arc from our given point, with the radius A 60, and upon that arc measure a chord equal to the radius, from point A to point 25; and lines drawn from these points to the given point will be at the required angle to each other.

FIG. 3.—Is a diagonal scale; it may usually be found engraved upon the reverse of Fig. 1. It is used to measure the decimal parts of an inch; the line A B is measured in inches, the first part, A O, being divided into ten equal parts, numbered from 0, 1, 2, 3, 4, &c., and a line A C, at a right angle to A B, which may be any length, is divided into ten equal parts, as 2, 4, 6, 8, and 10. The remaining lines being drawn as in the figure, it will be seen that the intersections of the vertical lines, with the oblique lines, gives us the divisions of hundredths of an inch, or tenths. For example, a measurement 2·27 of an inch would be from Z to Y, or 2·83 of an inch from X to V. These diagonal scales are sometimes made to show decimals of $\frac{1}{2}$ or $\frac{1}{4}$ inch, and other sub-divisions, as in a later part of this work.

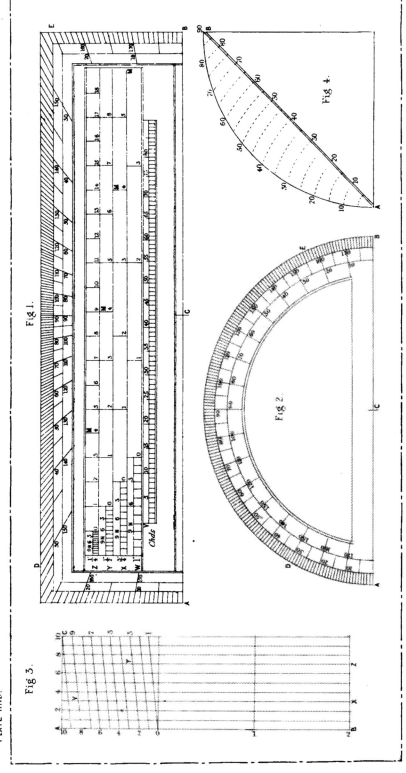

PLATE III.*a.*

Fig 1.

Fig 2.

Fig 3.

Fig 4.

Figs. 1 and 2 represent a pair of *set squares*, so called, from the fact of their always having two of their sides set at right angles, or what is commonly termed square, to each other. Fig. 1 has the two sides forming the right angle equal, and, as a consequence, the two other angles are equal, and each is equal to the half of a right angle, or 45°. In Fig. 2 one angle, (a), is made equal to 60°, which makes the remaining angle equal 30°.

These instruments are generally made of *wood*, but a substance of later invention, called *Vulcanite*, is much better, as it is less liable to variation from changes of atmosphere, and other causes. It must not be supposed that set squares are made only of these angles, but that these are the most useful to the student.

Figs. 3, 4, 5, 6.—These figures show the various uses to which set squares may be applied, either for making perpendiculars to given lines, or drawing parallels, or for bisecting a right angle; in all cases one square is held firmly in its place, while the second is first placed in the position of the solid lines, and then changed to the position of the dotted lines.

Fig. 7 shows how, by the use of the squares, a circle may be divided into 4, 6, or 12 equal parts; and similarly, the circle may be divided into 8 equal parts.

As he progresses through the earlier parts of this book, the student should practise his hands in the use of these squares, as he will find a great advantage from a facility in the manipulation of them, and a very great saving of time in all kinds of geometrical drawing from their dexterous use.

Fig. 8 shows a pair of common *bow pencil compasses*; these are the only instruments necessary for working out the figures, as far as pencil lines are concerned; but, if the figures are to be inked, he will require a pair of *bow pen compasses*, and a *ruling pen*. The working in ink is not, of course, requisite to the study of Geometry, but highly necessary in geometrical drawing generally.

Fig. 9 shows two views of the *shape* that the *lead point* in the compasses, and ruling pencil, should be cut to; this is called a *chisel point*, and should always be used edgewise, that a fine line may be drawn, and the point preserved for a longer time than by cutting in the ordinary way.

The earliest problems will show the student how lines may be drawn parl. and perpr. to each other, &c., by the use of a simple straight edge, and a pair of compasses, which are the only instruments allowed in a strict geometrical construction; but, since set squares are really substituted for such problems, it would be folly not to take advantage of such simple aids, which enable us to work the figures out so much more clearly and quickly.

PLATE III *b*.

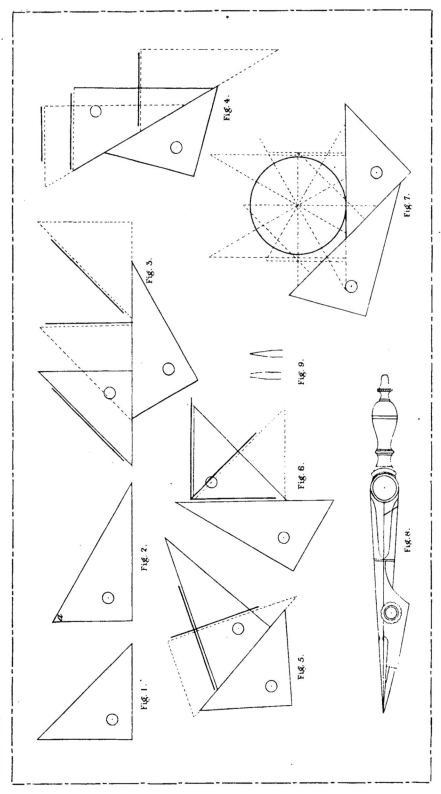

Fig. 1.

Fig. 2.

Fig. 3.

Fig. 4.

Fig. 5.

Fig. 6.

Fig. 7.

Fig. 8.

Fig. 9.

PART II.—FIRST SECTION.—RELATIONS OF LINES.—PLATE IV.

FIG. 1.—*To bisect a given line, as A B.**

From points A and B as centres, opening the compasses to any convenient radius greater than half A B, describe two arcs of the same radius intersecting in points C and D,—join points C and D by a right line, and it will bisect A B in point E. (See Note 4.)

FIG. 2.—*To draw a line perpendicular to a given line, from a given point in it, as A.**

From A, with any radius, mark off two points as B and C,—from B and C, with any radius greater than A B, describe two arcs intersecting in D,—draw a line from A through D, and it is the line required.

FIG. 3.—*To draw a line perpendicular to a given line, from a point without it, as A.**

From A, with any radius sufficiently great, describe an arc cutting the given line in B and C,—from B and C, with any radius greater than half C B, describe two arcs cutting in D,—draw a line from A through D, and it will be perpendicular to C B.

FIG. 4.—*To draw a line perpendicular to a given line, from a point nearly or quite at its extremity, as A.*

From any point as B, outside of the given line, rather nearer to A than to the opposite extremity, describe an arc through A, cutting the line in C, and produced beyond A,—from C draw a line through B, to cut the arc in D,—a line from D to A will be the line required.

Note 1.—The term line always implies a right line, unless otherwise mentioned, an arc of a circle always being expressed as an arc.

*** These three problems apply equally, and in exactly the same way, to an arc, as to a right line.

Note 2.—Arcs of circles are parallel when they are described from the same centre.

FIG. 5.—*To draw a line perpendicular to a given line, from a point nearly or quite over its extremity, as A.*

From A draw a line to any point in the line as B,—bisect B A in C,—from C, with CA as radius, describe an arc to cut the line in D,—draw A D.

FIG. 6.—*To draw a line parallel to a given line, as A B, at a distance from it equal to a given line, as C.* (Note 2.)

From any two points in the line A B, with line C as radius, describe two arcs on the same side of A B, and draw a line touching without cutting those arcs.

FIG. 7.—*To draw a line parallel to a given line, as A B, through a given point, as C.*

From C, with any sufficient radius, describe an arc to cut A B in D,—from D with the same radius describe an arc through C to cut A B in E,—make D F equal to E C,—and a line F C will be parallel to A B.

FIG. 8.—*To bisect a given angle, as B A C.* (Note 3.)

From the apex of the angle A, with any radius, describe an arc to cut the sides in points D and E,—from D and E, with any sufficient radius, describe two arcs cutting in F,—draw F A. (Note 5.)

Note 3.—To bisect, or trisect, in Geometry, always implies to cut in two, or three, equal parts.

Note 4.—The terms, to bisect by a perpendicular, will very frequently be used throughout this work, as signifying exactly what is done in Fig. I—i.e., the line is bisected by a line perpendicular to it; instead of saying, to bisect the line, and at the point of bisection to erect a perpendicular.

Note 5.—By the use of this problem, the student is enabled to find a great many angles, without using any protractor; as, for example, if he bisects a right angle, he has an angle of 45°, and if he adds the 45° to a right angle he will have an angle of 135°, and similarly a great many others may be found.

PLATE IV.

Fig 1.

Fig 2.

Fig 3.

Fig 4.

Fig 5.

Fig 6.

Fig 7.

Fig 8.

FIG. 9.—*To trisect a right angle, as* A B C.

From B, with any radius, describe an arc cutting the sides of the angle in points D and E,—from D and E, with the same radius, cut the arc in 1 and 2,—draw lines from B through 1 and 2.

FIG. 10.—*At a point in a given line, as* A, *to construct an angle equal to a given angle, as* B.

From the apex of the given angle, with any radius, describe an arc cutting its sides in points 1 and 2,—from A, with the same radius, describe an arc to cut the line in C,—make C D equal to 1 2,—draw a line from A through D.

FIG. 11.—*From a point outside a given line, as* A, *to draw a line making with the given line as* B C, *an angle equal to a given angle, as* X.

(fig.7)
(fig.10) Through A draw a line, A D, parl. to B C,—at A construct with the line A D, an angle equal to angle X, and produce the line down to cut B C in E,—A E B is equal to the given angle.

FIG. 12.—*To draw a line bisecting the angle formed by two lines, as* A B *and* C D, *without using the apex.*

(fig.6) At any distance, but equidistant from both, draw two lines, parallel to A B and C D, cutting in E,—from E, with any radius, describe an arc cutting A B and C D in F and G,—from F and G, with any radius, describe arcs intersecting in H,—a line drawn through H and E will be the line required.

FIG. 13.—*To draw a line through any given point, as* A, *to converge to the same point as two given lines, as* B C *and* D E, *when that point is inaccessible.*

(fig.6)
(fig.7) Draw any two lines parl. to each other, and cutting the given lines in points 1, 2, and 3, 4,—join 1 and 2 to point A,—draw lines from 3 and 4 parl. to lines 1 A and 2 A, intersecting in F,—a line through A and F is the line required.
Identically the same method may be pursued if the given point is outside the lines.

FIG. 14.—*Through a given point to draw a line to make equal angles with two given convergent lines,* A B *and* C D.

(fig.8) If the given point, as E, is on one of the lines, draw a line through it parl. to the opposite line, as E F,—bisect the angle A E F,—and the line bisecting it will, if produced, make equal angles with both lines,—if the given point is not in one line, two lines must be drawn through it parl. to the given lines.

FIG. 15.—*To draw a line through a given point, as* A, *to touch two given convergent lines, which line shall be bisected in* A.

Draw two lines from A, parl. to the given lines, to cut them in points B and C,—make C D equal to B A,—draw a line from D through A to E, it is bisected in A.

FIG. 16.—*To draw lines from any two given points, as* A *and* B, *to meet in, and to form equal angles with a given line,* C D.

(fig.3) From A draw a line perpr. to C D, cutting it in E, and produced beyond,—make E F equal to E A,—draw B F to cut C D in G,—join A G, and the problem is complete.
This one description applies equally to both cases, as in the figure.

PLATE V.

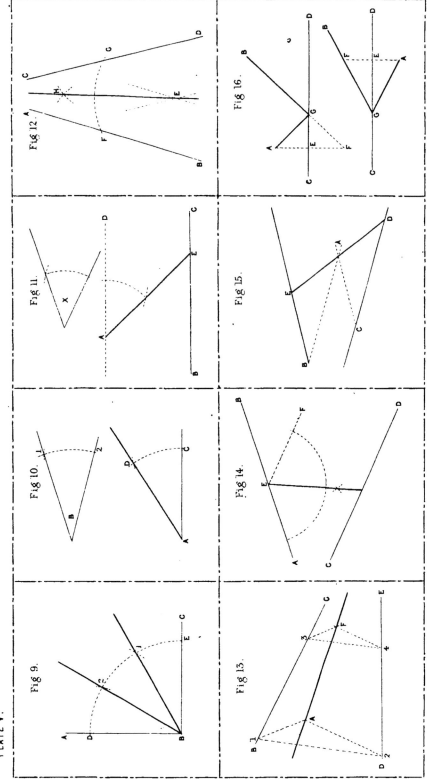

Fig 9.

Fig 10.

Fig 11.

Fig 12.

Fig 13.

Fig 14.

Fig 15.

Fig 16.

PART II.—FIRST SECTION.—RELATIONS OF LINES.—PLATE VI.

FIG. 17.—*To draw two equal lines from two given points, A and B, to meet in a given line.* (See Note 1.)

(fig.1) Draw A B, and bisect it by a perpr. produced to cut the given line in C,—join A C and C B.

FIG. 18.—*To divide a line, A B, into any number of equal parts, or segments.*

(fig.7) Draw a line, A C, from A, at any angle to A B,—draw a line from B parl. to A C,—from A and B step off on these lines a number of equal parts, equal, less one, to the number of parts A B is to be divided into, as 1, 2, 3, 4, 5, 6 and 7, to divide A B into eight equal parts, number them as in the figure,—join the corresponding points by lines which will divide A B into eight equal parts.

FIG. 19.—*A second method of the last problem.*

From A draw a line at any angle to A B, on it mark off from A, the same number of equal parts that it is required to divide A B into, as 1, 2, 3, 4 and 5,—join 5 and B,—draw lines from 1, 2, 3, and 4 parl. to 5 B, to cut A B, which will be divided as required. (Note 4.)

FIG. 20.—*To divide a line, A B, proportionally to a given divided line, C D.*

(fig.6) At any distance from A B draw a line, E F, equal to C D, and parl. to A B,—on E F mark points 1, 2, and 3, equal to the points on C D,—from A and B draw lines through E and F to meet in G,—from G draw lines through 1, 2, and 3 to cut A B, and they will divide it in parts in the same proportion to each other and to A B as the parts in the line C D.

Note 1.—The terms proportion, or ratio, signify the same thing; the ratio 1 to 2, or a to b, is just the same as the proportion of 1 to 2, or of a to b.

Note 2.—A line is divided externally in a given ratio, when the line has a portion added to it, such that the whole line so formed, bears the required ratio to the portion added.

Note 3.—A mean propnl., or as it is often written, a mean, between two lines, is a line, that bears the same ratio to the first of those lines, that the second bears to it; these lines are spoken of, as the two extremes and the mean.

FIG. 21.—*A second method of last problem.* (See Note 1.)

From A draw a line A E, at any angle to A B, equal to the given divided line, C D,—on A E mark, as in Fig. 20, the points or point F, as on C D,—join E and B,—draw F G parl. to E B.

FIG. 22.—*To divide a line, A B, externally, in a given ratio, as represented by lines C and D.* (See Note 2.)

From A and B draw two lines at any angle to A B, but parl. to each other,—make A E and B F equal respectively to C and D,—produce A B through B,—join E and F, producing the line to cut A B produced, in G,—B G will be in the same ratio to A G that the line D is to the line C, and thus it is said to be externally divided in the given ratio, while, in Fig. 21, A B is internally divided in the ratio of A F to F E.

FIG. 23.—*To divide a given line successively in its $\frac{1}{2}$, $\frac{1}{3}$, $\frac{1}{4}$, $\frac{1}{5}$, $\frac{1}{6}$, $\frac{1}{7}$, parts, &c.*

From A and B draw any two lines parl. to each other,—cut these lines in C and D, by any line parl. to A B,—draw C B and D A, cutting in 1,—from 1 draw a line parl. to A C, cutting A B in $\frac{1}{2}$,—draw line $\frac{1}{3}$ D, cutting C B in 2,—draw 2 $\frac{1}{3}$ parl. to A C,—from $\frac{1}{3}$ draw a line to D, cutting C B in 3, and proceed in this manner with point after point

FIG. 24.—*To find a mean propnl. between two given lines, as A B and C D.*

Upon A B, the greater of the two lines, describe a semicircle,—i.e., bisect A B in E, and from E, with radius E A, describe the semicircle,—from A or B mark off A F equal to C D,—from F draw a line perpr. to A B, to cut the semicircle in G,—G A is the mean proportional between C D and A B. (Note 3.)

Note 4.—The first of these methods is the more mathematical, but, if we take into consideration the use of set squares, this second method will be found to be in practice, immensely preferable.

The student, who has had no previous study of mathematics, will understand the latter part of this section more easily, after he has studied the other sections.

PLATE VI.

Fig 17.

Fig 18.

Fig 19.

Fig 20

Fig 21.

Fig 22.

Fig 23.

Fig 24.

FIG. 25.—*To divide a line, A B, internally, into extreme and mean proportion.* (See Note 2.)

Bisect A B in C,—at A or B erect a line perpr. to A B, and equal to BC, in D,—draw D A,—from D, with D B as radius, describe an arc cutting D A in E,—from A, with A E as radius, cut A B in F, and A B is divided in F, as required, for B F is in the same proportion to F A that F A is to A B. B F is one extreme, and F A the mean.

FIG. 26.—*To divide a line, A B, externally, into extreme and mean proportion.* (See Note 3.)

(fig.4) Bisect A B in C,—at A or B erect a perpr., B D, equal to A B,—from C, with C D as radius, describe an arc to cut A B, produced in E,—A E is now divided in extreme and mean, for E B is to A B, as A B is to A E.

FIG. 27.—*To find the Geometric, the Harmonic, and the Arithmetic, means, between any two given lines, as A B, and C D.* (See Note 1.)

Make E F equal to A B and C D, joined in G,—bisect E F in H,—from H, with H E as radius, describe an arc through E,—from G draw a perpr. to E F, to cut the arc in K,—G K is the geometric mean,—from G, with radius E H, cut the arc in L,—draw L G, producing it to cut the arc in M,—L G is the arithmetic, and G M is the harmonic mean between A B and C D.

FIG. 28.—*To find a third propnl. to two given lines, A B and C D, either greater or less.*

From A draw A E, equal to C D, at any angle to A B,—join

Note 1.—An Arithmetic mean, is a line, as much greater than one extreme, as it is less than the other.

An Harmonic mean, is a line, that bears the same relation to one extreme, that the other extreme, bears to the sum of the three lines together.

For Geometric mean, *see* Plate VI, Note 3.

E and B,—if a third propnl. less is required, from A as centre, describe an arc, E F,—draw F G parl. to B E,—G A is the third propnl. less than C D,—if a third greater is required, produce lines A E and A B,—make A H equal A B,—from H draw H K, parl. to E B, K A is a third propnl. to A B and C D, greater than either.

FIG. 29.—*To find a fourth propnl. either greater or less, or intermediate, to any three lines, A, B and C. First, a greater.*

Draw *a b* and *a c* at any angle to each other, and equal to lines B and C,—join *c b*,—produce *a c* and *a b*,—make *a e* equal line A,—draw a line from E parl. to *c b*, to cut *a b* produced in *f*,—*a f* is a fourth propnl. greater, for C : B :: A : *a f*; or *a c* : *a b* :: *a c* : *a f*.

FIG. 29A.—*Second, a less.*

Make lines Z Y and Z X equal line A and B,—join X and Y,—from Z, with radius equal line C, mark point W,—draw W V parl. to X Y,—V Z is the fourth propnl. less, for V Z : C :: B : A, or Z V : Z W :: Z X : Z Y.

FIG. 29B.—*Third, an intermediate, as between lines B and A.*

Draw P O and P Q equal to lines B and C,—join Q and O,—produce P O and P Q,—make P R equal line A,—draw R S parl. to O Q,—S P is a fourth propnl. between lines B and A, for C : B :: S P : A,—similarly, a propnl. may be found between lines B and C, by joining lines A and B, and marking line C upon B.

Note 2.—A line is divided medially, or in extreme and mean ratio, when it is so divided, that one part of the line, bears the same ratio to the whole line, that the second part bears to the first.

Note 3.—A line is divided medially, externally, when it is produced until the added portion, bears the same ratio to the original line, that the original line, bears to whole line so formed.

PLATE VII.

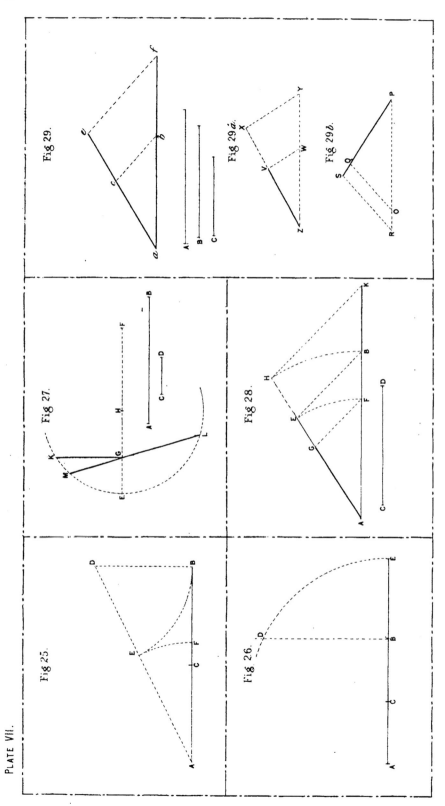

Fig 25.

Fig 26.

Fig 27.

Fig 28.

Fig 29.

Fig 29 a.

Fig 29 b.

FIG. 30.—*To divide a given line, A B, in such a point that the rectangle formed of its parts may be equal to a given square.*

Let C D represent the side of the given square,—on A B describe a semicircle,—at A erect a perpr., A E, equal C D,—from E draw a line parl. to A B, to cut the semicircle in F,—from F draw a perpr. to A B, to cut it in G,—the rectangle formed of the sides A G and G B will be equal to the square on C D.

FIG. 30A.—*Second case of last problem.*

Should C D be greater than half A B it may be divided externally, similarly, by joining E to A, and drawing a line from E perpr. to E A, to cut A B produced. This is virtually finding a third propnl. to lines A B and C D.

FIG. 31.—*To divide a line, A B, harmonically,* and in a given ratio, as Z and Y.*

Through A draw a line, at any angle, to A B,—make A C and A D each equal to the greater of the ratios, as Z,—from A make A E equal Y, the lesser ratio,—join C B,—from E draw a line parl. to A B, to cut C B in F,—from F draw a line parl. to C D, to cut A B in G,—from F draw a line to D, to cut A B in H,—points H and G, divide line A B harmonically, in the ratio of Z to Y, i.e., A B : G B : : G H : H A.

Note.—These problems will be more understood, by a student who has had no previous mathematical study, after he has studied Section V. of this Book.
Note. * A line is said to be harmonically divided, when it is divided in such a way,

FIG. 32.—*To find an harmonic mean between two given lines, as A B and C D.*

From one extremity of A B, as B, mark off C D in E,—
(fig.21) divide E A in H, in the ratio of A B to C D,—B H is then the mean required; for the harmonic progression is as C D or
B E : E H : : B A : H A.

FIG. 33.—*To find a third harmonic progressional to two given lines, as A B and C D.*

On A B make A E equal C D,—divide A B externally, in
(fig.22) the ratio of B E to E A in F,—A F is a third harmonic progressional.

FIG. 34.—*From a given point, A, to draw a line, to be cut by two given convergent lines, in segments, in a given ratio to each other.*

From A draw a line, A B, across the given lines,—divide
(fig.21) A B in the given ratio in C,—from C draw a line parl. to the given line farther from A, to cut the other given line in D,—A D continued to cut the farther line in E, is the line required.

Second Case.—When the point is between the given lines, draw a line through A, to cut one as A b,—make A c in the required
(fig.21) ratio to A b—from c draw a line parl. to the more distant line, to cut the nearer in d,—d c, drawn through A is the line sought.

in three parts; that either of the extreme segments, or parts, bears to the middle segment; the same ratio, that the other segment bears to the whole line.

To illustrate this, suppose a line is 12" long, and is divided into parts of 6", 2" and 4", then it is divided harmonically, for 4", one segment, is to 2", the second, as the whole line, 12", is to 6", the third segment.

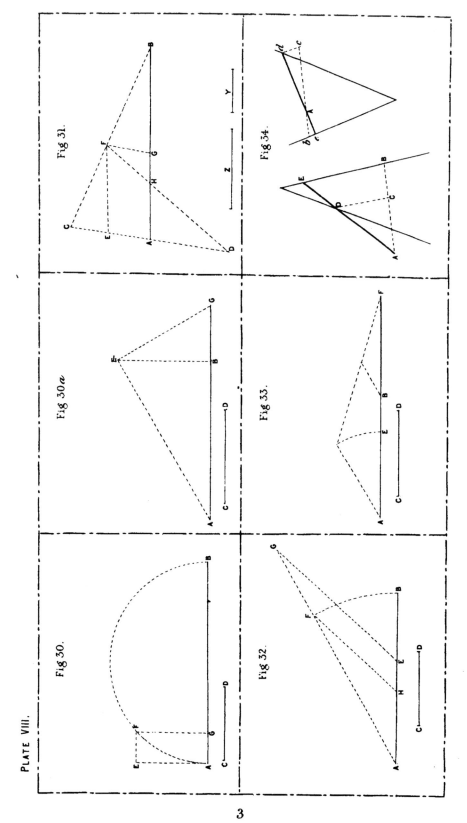

Fig 30.

Fig 30a

Fig 31.

Fig 32.

Fig 33.

Fig 34.

FIG. 35.—*Upon a given chord, A B, to describe a segment, to contain an angle equal to a given angle, C.*

(fig.10) At A construct an angle, B A D, equal to angle C,—bisect A B by a perpr., E F,—from A draw a line perpr. to A D, to cut E F in G,—from G, with G A as radius, describe the segment

FIG. 36.—*To cut off from a given circle, a segment, to contain a given angle.*

Draw a radius, A B,—at B draw a line, B C, perpr. to A B,—at B, on B C, construct an angle equal to the given angle, X,—produce the line to cut the circle in D,—and D B E is the segment required.

FIG. 37.—*To construct an equilateral triangle on a given line, A B.*

From A and B, with A B as radius, describe arcs intersecting in C,—draw C A and C B.

FIG. 38.—*To construct an equilateral triangle, its altitude, A B, being given.*

Through A and B draw lines perpr. to A B,—from A, with any radius, describe a semicircle, C D,—from C and D, with the same radius, mark off points E and F,—draw lines from A through E and F, producing them to cut the line through B in G and H,—triangle A G H will be eqrtl.

FIG. 39.—*To construct an isosceles triangle, its base, A B, and side, C, being given.*

From A and B, with the line C as radius, describe arcs cutting in D,—draw D A and D B.

FIG. 40.—*To construct an isosceles triangle, its altitude, A B, and vertical angle, C, being given.*

Through B draw a line perpr. to A B,—bisect angle C,—at A, on either side of A B, construct angles equal to the half of (fig.8) angle C,—and produce the sides of the angles to cut the line (fig.10) through B, in D and E.

FIG. 41.—*To construct an isosceles triangle, its base, A B, and its vertical angle, C D E, being given.*

Produce a side of the given angle to form the angle E D F, (fig.8) bisect angle E D F,—at A and B construct angles each equal (fig.10) to half angle E D F,—produce the sides of these angles to meet in G.

FIG. 42.—*To construct an isosceles triangle, its altitude, A B, and a base-angle, C, being given.*

Through A and B draw lines perpr. to A B,—at A construct angles with the line through A, each equal to angle C—produce the sides of these angles, to cut the line through B in D and E.

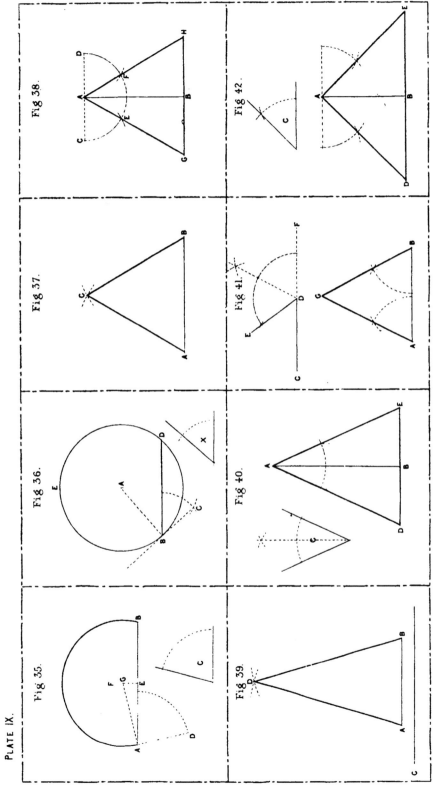

PLATE IX.

Fig 35.

Fig 36.

Fig 37.

Fig 38.

Fig 39.

Fig 40.

Fig 41.

Fig 42.

FIG. 43.—*On a given base, A B, to construct an isosceles triangle, the angle at whose base is double the angle at the vertex.*

(fig.26) Divide A B externally, in extreme and mean propn. in C,—from A and B as centres, with C A as radius, describe arcs intersecting in D,—join D A and D B.

FIG. 44.—*To construct an isosceles triangle of a given altitude, A B, and a given perimeter, C D.*

Bisect C D in E,—at E erect a perpr., E F, equal to A B,—
(fig.10) join F C and F D,—at F construct angles C F G and D F H, equal to the angles at C and D,—G F H is the triangle required.

FIG. 45.—*To construct a right-angled triangle, the hypothenuse, A B, and one acute angle, C, being given.*

At A or B construct an angle, D A B, equal to angle C,—
(fig.10) upon A B describe a semicircle cutting the line D A in E,—join E B.

FIG. 46.—*To construct a right angle triangle, its hypothenuse, A B, and one side, C D, being given.*

Upon A B describe a semicircle,—from A or B, with C D as radius, cut the semicircle in E,—draw E A and E B.

FIG. 47.—*To construct a right-angled triangle, its hypothenuse, A B, being given, and its sides being in proportion to each other.*

(fig.25) Divide A B, in extreme and mean propn., in point C,—upon A B describe a semicircle,—from C erect a perpr. to cut the semicircle in D,—join D to A and B,—D B, A D and A B are in continued proportion.*

FIG. 48.—*To construct a right-angled triangle, its hypothenuse, A B, being given, and the remaining sides being in a certain given ratio, as 5 : 3.*

Draw two lines, as Z Y and Z X, at right angles to each other,—from Z, with any radius, mark off five equal parts on one line, and, with the same radius, three equal parts on the second line,—draw a line through points 3 and 5,—on line 3 5, or 3 5 produced through 5, set 3 V, equal to A B,—from V draw a line parl. to X Z, to cut Y Z, or Y Z produced, in W.—V 3 W is the required triangle.

FIG. 49.—*To construct a scalene triangle, its three sides, A B, C D, and E F being given.*

From A, with C D as radius, describe an arc, and from B, with E F as radius, describe a second arc, cutting the first in G,—join G A and G B.

FIG. 50.—*To construct a triangle, its base, A B, and the angles C and D at the base, being given.*

At A construct an angle equal to the angle C,—at B construct an angle equal to the angle D,—produce the sides of the angles to meet in E.

Note.—The hypothenuse is always the side opposite the right angle.
* The earlier problems here used, are worked in the dotted lines, as an aid to the student, but they are not re-described; the same will be frequently found in the figures, without particular reference.

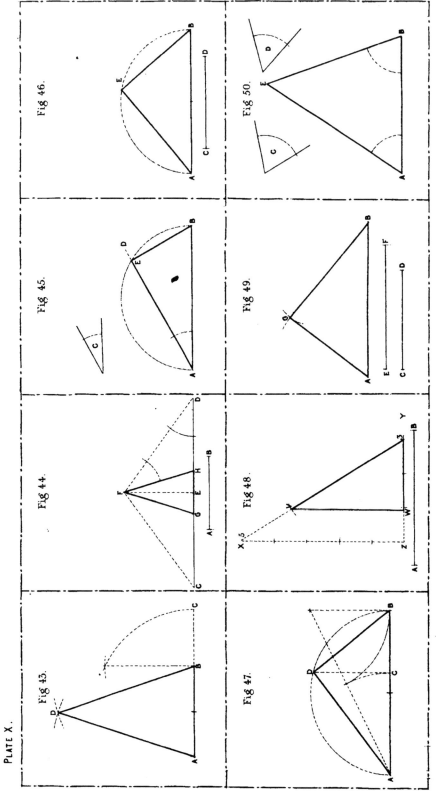

PLATE X.

Fig 43.

Fig 44.

Fig 45.

Fig 46.

Fig 47.

Fig 48.

Fig 49.

Fig 50.

FIG. 51.—*To construct a triangle, the altitude, A B, and the angle C, at the vertex, and the angle D, at the base, being given.*

Through A and B draw lines perpr. to A B,—at A construct an angle, E A F, equal to angle D,—at A construct an angle, F A G, equal to the angle C,—produce A F and A G to meet the line through B in I and H.

FIG. 52.—*To construct a triangle, the base, A B, the angle C, at the vertex, and one angle, D, at the base, being given.*

At A construct an angle, E A B, equal to angle C,—from B draw a line parl. to A F, to cut A E produced in G,—A B G is the triangle required.
(fig.7)

FIG. 53.—*To construct a triangle having its sides in a given ratio, as 4, 5, 6, and its perimeter, A B, being given.*

Divide A B into 15 (the sum of the three sides) parts,—from point 4, with 4 A as radius, describe an arc,—from point 9, with 9 B as radius, describe a second arc, to cut the first in C,—draw C 4 and C 9.

FIG. 54.—*To construct a triangle, its base, A B, its altitude, C D, and one side, E F, being given.*

At a distance from A B equal to C D, draw a line parl. to A B,—from A or B, with E F as radius, describe an arc cutting the parl. in G,—draw G A, and G B.
(fig.6)

FIG. 55.—*To construct a triangle, its base, A B, its altitude, C D, and one angle, E, at the base, being given.*

At a distance from A B equal to C D, draw a line parl. to A B,—at A or B construct an angle equal to the given angle, E, producing the side of the angle to cut the parl. in F,—draw F A.
(fig.6)

FIG. 56.—*To construct a triangle, its base, A B, one side, C D, and its perimeter, E F, being given.*

Make E G equal C D, and G H equal A B,—from G, with radius G E, and from H, with radius H F, describe arcs intersecting in I,—draw I G and I H.

FIG. 57.—*To construct a triangle, its perimeter, A B, and two of its angles, C and D, being given.*

At A construct an angle equal to angle C,—at B construct an angle equal to angle D, producing its side to meet the line from A, in E,—bisect the angles E A B and E B A by lines meeting in F (this point F, which is equidistant from all the sides of the triangle E A B, is its centre)—from F draw lines parl. to E A and E B, to cut A B in G and H,—F G H is the triangle required.
(fig.7)

FIG. 58.—*To construct a triangle, its base, A B, its altitude, C D, and the angle E, opposite its base, being given.*

On A B, as a chord, describe a segment, to contain an angle equal the given angle, E,—at a distance from A B equal to C D, draw a line parl. to A B, to cut the segment in F or G,—join F or G to A and B.
(fig.35)
(fig.6)

PLATE XI

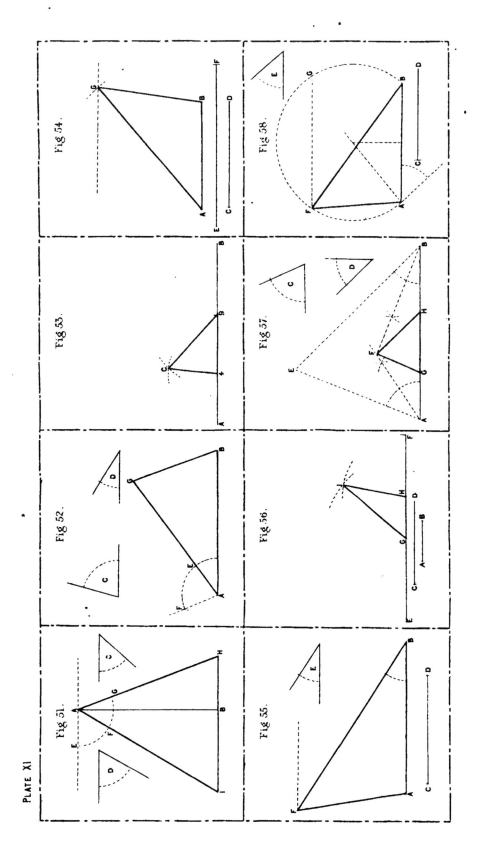

Fig 51.   Fig 52.   Fig 53.   Fig 54.

Fig 55.   Fig 56.   Fig 57.   Fig 58.

Fig. 59.—*To construct a triangle, one angle, A, the side opposite to it, B C, and the sum of the sides containing it, D E, given.*

At D construct an angle, F D E, equal to A,—bisect F D E by D G,—from E, with radius B C, cut D G in H,—draw H I parl. to D F,—join H E.

Fig. 60.—*To construct a triangle, one angle, A, the side adjacent to it, B C, and the difference between the remaining sides, D E, being given.*

At B construct an angle, F B C, equal to angle A,—on F B make B G, equal to D E,—join G C,—at C construct an angle on C G equal to angle F G C,—produce the side to meet F G produced in H,—H B O is the triangle.

Fig. 61.—*To construct a triangle, one angle, A, the side opposite to it, B C, and the difference between the remaining sides, D E, being given.*

At one end of D E erect a perpr., as E F,—bisect angle A,—at E make angle F E G, equal half angle A,—from D, with radius B C, cut E G in H,—bisect H E by a perpr, produced to cut D E produced in I,—H D I is the triangle.

(Note 4, Plate IV.)

Fig. 62.—*To construct a triangle, having its base, A B, one angle, C, and the sum of the remaining sides, D E, given.*

At A, construct on A B an angle, equal to the angle C,—make the side A F equal to D E,—join F B,—bisect F B by a perpr. to cut F A in G,—join G B,—A G B is the triangle.

Fig. 63.—*To construct a triangle, its base, A B, the opposite angle, C, and the sum of the remaining sides, D E, given.*

Bisect angle C,—on A B describe a segment to contain an (fig. 35) angle equal to half angle C,—from A, with radius D E, cut the segment in F,—join F B and F A,—bisect F B and F A,—by a perpr. to cut F A in G,—join G B,—A B G is the triangle.

Fig. 64.—*To construct a triangle, the altitude, A B, the perimeter, C D, and the base angle, E, being given.*

At a distance from C D equal to A B, draw a line, F G, parl. to C D,—at C construct an angle equal angle E, to cut F G in H,—make C I equal C H,—from I draw a line parl. to C H, to cut F G in K,—join K D,—bisect K D by a perpr. to cut C D in L,—I K L is the triangle.

Fig. 65.—*To construct a triangle, its perimeter, A B, its altitude, C D, and the angle at the apex, E, being given.*

At A construct an angle, F A B, equal angle E,—bisect F A B by G A,—bisect A B by a perpr. to cut G A in H,—from H, with radius H A, describe an arc, A B,—at a distance from A B equal to C D, draw a line parl. to it, to cut the arc in a point, as I,—join I A and I B,—bisect I A and I B by perprs, to cut A B in K and L,—join I K, I L,—K I L is the required triangle.

Fig. 66.—*To construct a triangle, its base, A B, and its altitude, C D, being given, with its sides in a given ratio, as line Z : X.*

(fig. 21)
(fig. 22)

Divide A B internally, in the given ratio, in E,—divide A B, externally, in the given ratio, in F,—upon E F describe a semi-circle; from any point in this semicircle lines drawn to A and B will be in the given ratio,—at a distance from A B equal C D draw a line parl. to A B, to cut the semicircle in G or H,—join either G or H to points A and B,—triangles A G B or A H B both fulfil the conditions given.

PLATE XII.

Fig 59.

Fig 60.

Fig 61.

Fig 62.

Fig 63.

Fig 64.

Fig 65.

Fig 66.

FIG. 67.—*To construct a square, one side, A B, being given.*

At A or B erect a perpr. to A B,—make A C equal A B,—from C and B, with A B as radius, describe two arcs intersecting in D,—draw D C and D B.

FIG. 68.—*To construct a square, one diagonal, A B, being given.*

Bisect A B by a perpr. in C,—make C D and C E each equal to A C,—join D and E to A and B.

FIG. 69.—*To construct an oblong or rectangle, its two different sides being given, as A B and C D.*

At A erect a perpr. to A B,—make A E equal to C D,—from E, with A B as radius, and from B, with C D as radius, describe arcs cutting in F,—draw F B, F E.

FIG. 70.—*To construct an oblong, one diagonal, A B, and one side, C D, being given.*

Bisect A B in E,—from E, with radius E A, describe a circle about A B,—from A and B, with radius C D, cut the circle on opposite sides of A B, in G and F,—join F and G to A and B.

FIG. 71.—*To construct an oblong, its diagonal, A B, and the angle O made by the side with the diagonal, being given.*

As in the last figure, describe a circle about A B,—at A construct an angle equal to the angle O, producing the side to cut the circle in D,—draw B E parl. to A D, and join A E and D B.

FIG. 72.—*To construct an oblong, its diagonal, A B, being given, and the sides in a certain ratio or proportion, as 3 and 2.*

As in Fig. 66, divide A B internally and externally, in the given ratio, in C and D,—and describe a semicircle upon C D,—about A B describe a circle to cut the semicircle in E,—draw B F parl. to A E,—join E and F to B and A.

FIG. 73.—*To construct a rhombus, one side, A B, and one angle, C, being given.*

At A construct an angle equal C,—make A D equal A B,—from D and B, with radius A B, describe arcs intersecting in E,—draw D E and B E.

FIG. 74.—*To construct a rhombus, one diagonal, A B, and one side, C D, being given.*

From A and B, with radius C D, describe arcs cutting in points E and F,—join the four points A, E, B, and F.

PLATE XIII.

Fig 67.

Fig 68.

Fig 69.

Fig 70.

Fig 71

Fig 72.

Fig 73.

Fig 74.

FIG. 75.—*To construct a rhombus, its diagonals, A B and C D, being given.*

Bisect A B by a perpr. in E,—bisect C D in F,—make E G and E H each equal F C,—join the four points A, G, B, and H.

FIG. 76.—*To construct a rhomboid, two unequal sides, A B and C D, and one angle, E, being given.*

At A construct an angle equal to angle E,—make A F equal C D,—from point F, with radius A B, and from B, with radius C D, describe arcs to cut in G,—draw G F and G B.

FIG. 77.—*To construct a rhomboid, one diagonal, A B, and the two different sides, C D and E F, being given.*

From A and B, with radius C D, describe arcs on the opposite sides of A B,—from A and B, with radius E F, describe arcs to cut the first arcs in points G and H,—join the four points A, G, B, and H.

FIG. 78.—*To construct a rhomboid, its two diagonals, A B and C D, and the angle made by them, E, being given.*

Bisect A B in F, and C D in G,—through F draw a line, making with A B an angle equal angle E,—from point F, with radius C G, mark off on it points H and I,—join the four points A, H, B and I.

FIG. 79.—*To construct a rhomboid, its two diagonals, A B and C D, and one side, E F, being given.*

Bisect A B in G and C D in H,—from G, with radius C H, describe a circle,—from A and B, with radius E F, describe arcs on opposite sides of A B, cutting the circle in points I and K,—join the four points A, I, B, and K.

FIG. 80.—*To construct a trapezium equal to a given trapezium, A B C D.*

Draw a line, E F, equal line A B,—at E construct an angle equal C A B,—make E G equal A C,—from G, with radius C D, and from F, with radius B D, describe arcs cutting in H,—draw H G and H F.

FIG. 81.—*On a given line, E F, to construct a trapezium similar to a given trapezium, A B C D.*

At E make angle G E F equal angle C A B,—draw C B,—at F construct an angle equal angle C B A, producing the side to cut G E in H,—draw D A,—at E construct an angle, I E F, equal D A B,—at H construct an angle with H E equal D C A,—produce the side to cut E I in K,—join K F.

FIG. 82.—*To construct a trapezion,* * its two diagonals, A B and C D, and one side, E F, being given.*

(* Def. 25.)

Bisect C D in G,—on both sides of A B, at a distance equal C G, draw lines parl. to A B,—from A or B, with radius E F, describe an arc to cut the parls. in H and I,—join points A, H, B, and I.

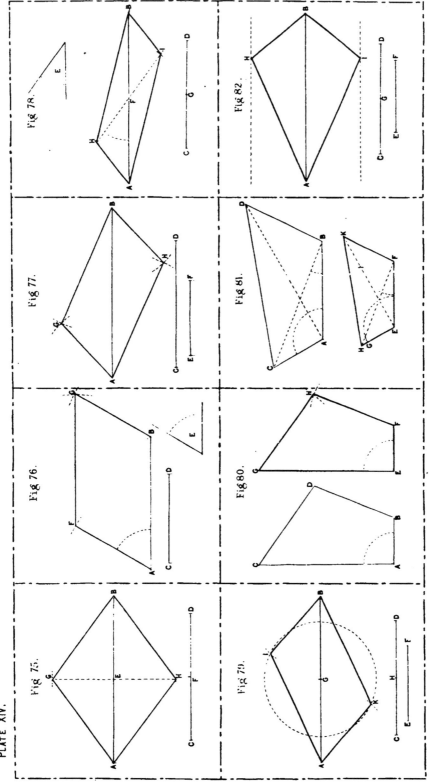

PLATE XIV.

Fig. 75.

Fig. 76.

Fig. 77.

Fig. 78.

Fig. 79.

Fig. 80.

Fig. 81.

Fig. 82.

FIG. 83.—*To construct a trapezium, one side, A B, one diagonal, C D, and one angle, E, being given.*

At A construct an angle equal E,—make A F equal A B,—bisect angle F A B, making A G equal line C D,—join point G to F and B.

FIG. 84.—*To construct a trapezium, when two unequal sides, A B and C D, and one angle, E, are given.*

At A construct an angle equal E,—make A F equal A B,—from B and F, with radius C D, describe two arcs cutting in G,—draw lines G F and G B.

FIG. 85.—*To construct a trapezium, with sides in a given ratio, as 2, 3, 4, and 5, the perimeter, A B, and one angle, C, given.*

Divide A B in the given ratios, as 2, 3, 4, and 5, by dividing it into fourteen parts (the sum of the numbers of the ratios), and marking off 2nd, 5th, and 9th of these parts,—at B construct an angle equal angle C,—make B D equal line A 2,—from D, with radius line 2, 5, and from point 9, with radius 5, 9, describe arcs intersecting in E,—draw E D and E 9.

FIG. 86.—*To construct a trapezium, with sides in a given ratio, as 3, 4, 5, 6, when one side, A B, and one angle, C, are given.*

Divide A B into a number of equal parts, equal to one of the numbers of the ratio, as 4,—produce A B, marking on it the same equal parts to the highest number of the given ratio, as 6,—at A construct an angle equal to angle C,—make A D equal three of the equal parts,—from D, with radius equal five of the parts, and from B, with radius equal six of the parts, describe arcs cutting in E,—draw lines E D and E B.

FIG. 87.—*To construct any regular polygon, its circumscribing circle being given.*

Draw a diameter to the given circle, as A B,—divide A B into the same number of equal parts as 7, that the polygon is to have sides,—from A and B as centres, with A B as radius, describe arcs cutting in C,—from C draw a line through the second division, from either end of the diameter, producing it to cut the circle in D,—the distance from D to the nearest end of the diameter is one side of the required polygon,—mark the radius, A D, off round the circle from D, in points 1, 2, 3, 4 and 5,—join all these points and A.

FIG. 88.—*To construct any regular polygon, one of its sides, A B, being given.*

At A erect a perpr. to A B,—from A, with radius A B, describe a quadrant, B C,—divide the quadrant B C, by trials with the compasses, into the same number of equal parts that the polygon is required to have sides,—from A draw a line through the second point of division from C,—bisect A B by a perpr., producing it to cut the line from A through 2 in D,—point D is the centre of the required polygon,—from D, with radius D A, describe a circle through points A and B,—from A or B set off the side A B on the circle in points 1, 2, 3, &c, and connect these points together with A and B.

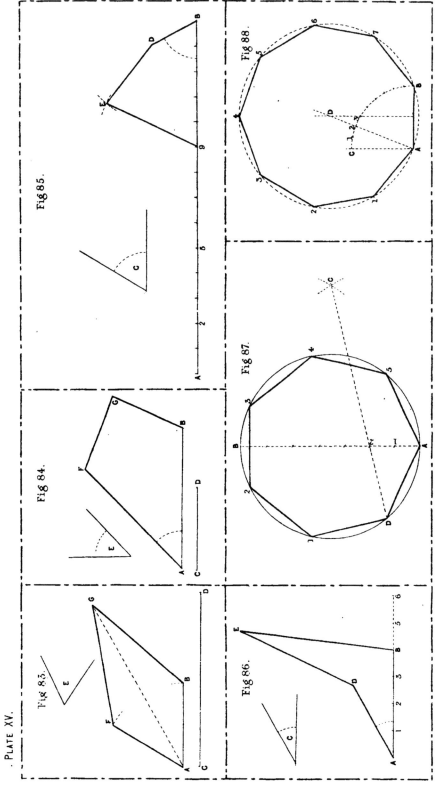

PLATE XV.

Fig 83.

Fig 84.

Fig 85.

Fig 86.

Fig 87.

Fig 88.

FIG. 89.—*To construct any regular polygon up to 12 sides, one side, A B, being given.*

Bisect **A B** by a perpr. from **C**,—from **A**, with radius **A B**, describe an arc from **B**, to cut the perpr. in **D**,—divide arc **B D** into six equal parts by bisecting it, and then dividing each half into three equal parts by trials,—number the points of division 1, 2, 3, 4, 5, from the point **D**,—point **D** is the centre of a hexagon, of which **A B** is one side,—produce **C D** through **D**,—from **D**, with radius equal **D 1**, mark on line **C D** points 5 and 7,—point 5 is the centre of a pentagon, of which **A B** is one side,—point 7 is the centre of a heptagon, **A B** being one of its sides, which may be said of all the following centres.

From **D** as centre, with chord **D 2** as radius, cut **C D** in point 8,—point 8 is the centre of an octagon,—from **D**, with **D 3** as radius, mark point 9 on **D C**,—point 9 is the centre of a nonagon,—from **D**, with **D 4** as radius, mark point 10, it is the centre of a decagon,—from **D**, with **D 5** as radius, mark point 11,—it is the centre of an undecagon,—and marking point 12, from **D**, with **D B** as radius, we have the centre of a duodecagon.

Having these centres, and the line **A B** as one side, to complete any one of the polygons, we must describe a circle from the centre, with the distance to **A** or **B** as radius, and upon this circle, mark off the line **A B**, all round, from **A** or **B**, and we shall have the points of the polygon, which must be joined by right lines.

FIG. 90.—*To construct a regular polygon, two sides, A B and A C, and the angle C A B contained by them, being given.*

Bisect **A B** and **A C** by perprs., producing them to meet in point **D**,—**D** is the centre of the polygon,—from **D**, with radius **D A** or **D B**, describe a circle passing through **B** and **A**,—from **B** and **C** set off on the circle the radius **A B**, and join all the points so obtained.

FIG. 91.—*To construct a pentagon, one side, A B, being given.*
(fig. 26)

Divide **A B** medially, externally, in point **C**,—from **A** and **B**, with **A C** as radius, describe two arcs cutting in **D**,—from **D**, with **A B** as radius, describe an arc towards **A B**,—and from **A** and **B**, with the same radius, cut this arc in points **E** and **F**,—join the points **A**, **E**, **D**, **F**, **B**.

FIG. 92.—*To construct a pentagon, one diagonal, A B, being given.*
(fig. 25)

Divide **A B** medially, internally, in point **C**,—from **A** and **B**, with **A B** as radius, describe arcs through **A** and **B**,—from **A** and **B**, with **A C** as radius, mark on these arcs points **D** and **E**, and on the opposite side of **A B** describe other arcs, intersecting in **F**,—join the five points **A**, **F**, **B**, **E**, **D**.

FIG. 93.—*To construct a hexagon on a given line, A B.*

From **A** and **B** as centres, with radius **A B**, describe arcs intersecting in **C**, and extended outwards from **C**,—from **C**, with the same radius, describe a semicircle cutting these arcs in **D** and **E**,—from **D** and **E**, with the same radius, cut the semicircle in **F** and **G**,—join the six points **A**, **E**, **F**, **G**, **D**, and **B**.

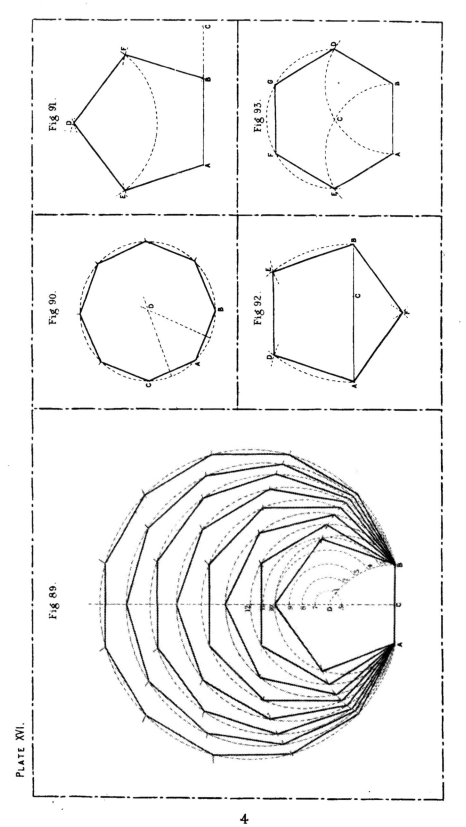

PLATE XVI.

Fig 89.

Fig 90.

Fig 91.

Fig 92.

Fig 93.

4

Fig. 94.—*To construct a hexagon, its long diagonal, A B, being given.*

Bisect A B in C,—from C, with C A as radius, describe a circle,—from A and B as centres, with the same radius, cut the circle on both sides of A B, in points D E F and G,—join the six points on the circle.

Fig. 95.—*To construct a hexagon, its short diagonal, A B, being given.*

On A B construct an eqrtl. triangle, A B C,—bisect the angles of the triangle by lines A D, B E, and C F, crossing in X,—from X, with the distance from X to A as radius, describe a circle to cut the lines A D, B E, and C F, in points 1, 2, and 3,—join these points to the points A B and C.

Fig. 96.—*To construct a hexagon, its diameter, A B, being given.*

Bisect A B in C by a perpr.,—from C, with any radius, describe a quadrant, D E,—from D, with the same radius, mark point F,—at B erect a perpr. to A B,—draw line C F to cut the perpr. through B in G,—from C, with the radius C G, describe a circle,—from G, with the same radius, mark on the circle points 1, 2, 3, 4, and 5,—join all these points and the point G.

Fig. 97.—*To construct a heptagon, its circumscribing circle being given.*

Draw a radius, A B,—bisect A B in C by a perpr. to cut the circle in D,—line C D equals one side of the heptagon,—set line C D off round the circle and join the points.

Fig. 98.—*To construct a heptagon, its longer diagonal, A B, being given.*

From A and B, with the radius A B, describe arcs to cut in C,—from C draw a line, C D, perpr. to A B,—from A, with radius C D, cut arc A C in E,—draw B E cutting C D in F, bisect arc E A in G,—make B H equal A G,—from G and H, with radius A B, describe arcs through A and B,—from A and B, with radius A G or B H, describe arcs to cut the last in points I and K,—join the points A, G, F, H, B, K, I and A.

Fig. 99.—*To construct a heptagon, its shorter diagonal, A B, being given.*

From A and B, with A B as radius, describe arcs crossing in C,—draw C D perpr. to A B,—from A and B, with radius C D, cut arcs A C and B C in points E and F,—from A and B draw lines through E and F, producing them to cut the arcs in points G and H,—join G and H,—from points A and B, A and G, and H and B, describe arcs with radius G H cutting in points I, K, and L,—join these last points to points A B G and H.

Fig. 100.—*To construct an octagon, its circumscribing circle being given.*

Draw two diameters perpr. to each other, cutting the circle in four quadrants, in points 1, 2, 3, and 4,—bisect these quadrants in points 5, 6, 7, and 8,—join the eight points.

Fig. 101.—*To construct an octagon, on a given line, A B.*

At A and B erect lines A C and B D, perpr. to A B,—from A and B, with radius A B, cut these lines in E and F,—draw lines A F and B E, producing them beyond E and F,—make E G equal A B, and F H equal E G,—from G and H, with the same radius, cut lines A C and B D in points I and K,—from G and A, and from B and H, with radius A B, describe arcs cutting in L and M,—join points A, L, G, I, K, H, M and B.

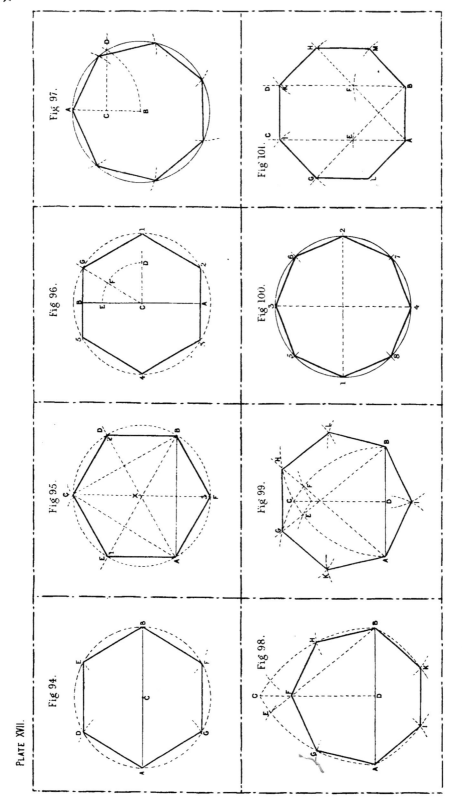

PLATE XVII.

Fig 94.

Fig 95.

Fig 96.

Fig 97.

Fig 98.

Fig 99.

Fig 100.

Fig 101.

FIG. 102.—*To construct an octagon, its longest diagonal, A B, being given.*

Bisect A B in C by a perpr.,—from C, with C A as radius, describe a circle cutting the perpr. through C in D and E,—bisect the arcs A D, D B, B E, and E A, and join the points of bisection to points A, D, B, and E.

FIG. 103.—*To construct an octagon, its medium diagonal, A B, being given.*

At A and B erect perps. to A B, as A C and B D,—bisect angle D B A by line B E,—bisect angle E B A by a line cutting A C in F,—bisect F B in G,—from G, with radius G F, describe a circle,—from B and A, with radius A F, mark off on the circle the remaining points of the octagon, H, I, and K.

FIG. 104.—*To construct an octagon, its shortest diagonal, A B, being given.*

Upon A B construct a square, A C D B,—draw the diagonals A D and C B, crossing in E,—from E, with E A as radius, describe the circle,—bisect the arcs A C, C D, D B, and B A, in points 1, 2, 3, and 4,—join the eight points in the circle.

FIG. 105.—*To construct an octagon, its diameter, A B, being given.*

Bisect A B in C,—through A and B erect perps. to A B,—make A D, A E, B F and B G each equal to A C,—join D F and E G,—from points D, F, G, and E, with radius D C, describe arcs cutting these lines in points 1, 2, 3, 4, 5, 6, 7, and 8,—join these points for the octagon.

FIG. 106.—*To construct a duodecagon, its longest diagonal, A B, being given.*

Bisect A B in C,—from C, with C A as radius, describe a circle,—from A and B, with the same radius, mark points D E F and G,—bisect arcs A D, D E, E B, B G, G F, and F A,—join all the points on the circle.

FIG. 107.—*To construct a duodecagon, its second longest diagonal, A B, being given.*

On A B construct an eqrtl. triangle, A B C,—bisect the angles of the triangle by lines crossing in D,—from D, with D A as radius, describe a circle cutting these lines in points E, F, and G,—bisect the arcs A E, E C, C F, F B, B G, and G A,—join the points in the circle.

FIG. 108.—*To construct a duodecagon, its second shortest diagonal, A B, being given.*

On A B construct a square, A C D B,—draw the diagonals, A D and C B, cutting in E,—from E, with radius E A, describe a circle,—from A, C, D, and B, with radius A E, mark on the circle points 1, 2, 3, 4, 5, 6, 7, and 8,—join all the points on the circle.

FIG. 109.—*To construct a duodecagon, its shortest diagonal, A B, being given.*

From A and B, with A B as radius, describe arcs intersecting in C,—from C, with radius C A, describe a circle,—from A or B, with the same radius, mark on the circle points D E F and G,—bisect the arcs A B, A D, D E, E F, F G, and G B,—join all the points on the circle.

PLATE XVIII.

Fig 102.

Fig 103.

Fig 104.

Fig 105.

Fig 106.

Fig 107.

Fig 108.

Fig 109.

Fig. 110.—*To construct an irregular pentagon, with the following conditions:*

sides A B equal 2"		angles A B C equal 120°	
" B C	= 2·3",	" B C D	= 110°
" C D	= 2·6",	" C D E	= 100°
" D E	= 2·7",		—join E and A.

(The measurements in the plate are a ¼ size.

Draw A B equal 2",—at B construct an angle, A B C, equal 120°,—make B C equal 2·3",—at C construct an angle, B C D, equal 110°,—make C D equal 2·6",—at D construct an angle, C D E, equal 100°,—make D E equal 2·7",—join E and A. (The measurements in the plate are a ¼ size.

Fig. 111.—*To construct an irregular hexagon, with the following conditions:*

sides A B equal 2"		diagonals A C	= 3"
" B C	= 1¾"	" B F	= 2½"
" A F	= 1¾"	" B E	= 3½"
" F E	= 2½"		
" A D	= 2½"		
" C D	= 1"		

Make A B equal 2",—from A, with radius 3", and from B, with radius 1¾", describe arcs cutting in C,—from A, with radius 1¾", and from B, with radius 2½", and from F, with radius 2½", describe arcs cutting in F,—from B, with radius 3½", describe arcs cutting in E,—from E, with radius 1¾", describe arcs cutting in E,—from C, with radius 1", describe arcs cutting in D,—join points A, F, E, D, C, and B. (This is worked half the size, on the plate.)

Fig. 112.—*To construct an irregular heptagon, under the following conditions:*

diagonals B G equal 1·4"		sides A B	= ·8"
" A C	= 1·4"	" B C	= ·75"
" B D	= 1·6"	" A G	= ·75"
" A F	= 1·6"	angles A G F	= 120°
" A E	= 1·2"	" B C D	= 120°
" B E	= 1·2"		

Make A B equal ·8",—from A, with radius 1·4", and from B, with radius ·75", describe arcs cutting in G,—from B, with radius 1·4", and from A, with radius ·75", describe arcs cutting in G,—at G and C construct

angles A G F and B C D, each equal 120°,—from A, with radius 1·6", mark point F, and from B, with radius 1·6", mark point D,—from A and B, with radius 1·2", describe arcs cutting in E,—join points F E and E D.

Fig. 113.—*To construct an irregular octagon under the following conditions:*

Its shortest diagonal, ·75" long, subtends an angle of 90°, and joins two angles of 165° each, its opposite sides are parl., and all the sides equal.

Draw A C equal ·75" as the shortest diagonal,—bisect A C in X, by a perpr.,—make X B equal A X,—at C, on B C, construct an angle, B C D, of 165°,—make C D equal B C,—at A construct an angle, B A H, equal 165°,—make A H equal A B,—from D draw D E, parl. and equal A H,—from H draw H G, parl. and equal to C D,—from E and G, with E D as radius, describe arcs cutting in F,—draw G F and F E.

Fig. 114.—*To construct a figure of a given perimeter, Z Y, similar to a given figure, A B C D E.*

Produce the base, A E,—on the production from E mark the sides E D, D C, C B, and B A, in points 1, 2, 3, and 4,—draw Z Y anywhere that is convenient, parl. to A E,—from A draw a line through point Z, and from point 4 draw a line through Y, producing it to cut line A Z in X,—from the divisions E, 1, 2, and 3, draw lines to X, cutting Z Y in V, U, T, and S,—from V draw V R parl. to E D, and equal V U,—from R draw R Q, parl. to D C, and equal to U T,—from Q draw Q P, parl. to C B, and equal T S,—join P and Z.

Fig. 115.—*On a given base, or base produced, as A Z, to construct a figure similar to a given figure, A B C D E F.*

From A draw lines through the other points of the figure producing them outwards,—draw a line from Z parl. to B C, to cut A C produced in Y,—from Y draw a line parl. to C D, to cut A D produced in X,—from X draw a line parl. to D E, to cut A E produced in W,—from W draw a line parl. to E F, to cut A F produced in V. (Should the given base be less than A B, the figure is similarly constructed within the given figure.

Note.—These examples of irregular polygons are given, not as embodying any principles of construction, but as exercises, to suggest to the student, a few of the many varieties of forms, in which such problems may be put.

PLATE XIX.

Fig 110.

Fig 111.

Fig 112.

Fig 113.

Fig 114.

Fig 115.

FIG. 116.—*To find the curve of an ellipse, by points, its major and minor axes, A B and C D, being given.*

Place A B and C D at right angles to, and bisecting each other in E,—from E, with radii E A and E C, describe two circles,—mark any number of corresponding divisions on these circles, by drawing lines from the centre, to cut both the circles in points 1, 1, 2, 2, 3, 3, &c.,—from the points in the larger circle, draw lines parl. to the minor axis, and from the points in the lesser circle, draw lines parl. to the major axis, the lines from corresponding points, crossing in points o, o, o, &c., through which points, the curve of the ellipse may be drawn by hand, or with a French curve. (The operation must be repeated in each quarter.)

FIG. 117.—*To find the curve of an ellipse by points, two conjugate diameters, A B and C D, and the angle formed by them, being given.* (See Note 1.)

Through points A, B, C, and D, draw lines parl. to A B and C D, enclosing the diameters in a pgram, E F H G,—divide the half of A B into any number of parts,—divide F B and B H into the same number of parts, (fig. 21) in the same ratio as the half A B,—number these points as 1, 2, 3, &c., both ways, from point B,—through the points on B F draw lines from point C,—from point D draw lines through the corresponding points on B A, to cut the first lines in points, as X X X, &c.,—through these points draw the curve, and repeat the work in each quarter. (See Note 2.)

FIG. 118.—*To find the approximate curve of an ellipse by arcs of circles, the axes, A B and C D, being given.*

Place A B and C D at right angles to, and bisecting each other,—from B, with radius C D, mark point E,—trisect line E A,—from centre, X, with A 2 as radius, mark on either side, points F and G,—from F and G, with F G as radius, describe arcs on both sides of A B, cutting in H and I,—from H and I draw lines through points F and G, producing them beyond,—from H, with H D as radius, describe arc K L,—from I, with I O as radius, describe arc M N,—from points F and G, with radius F A, describe the arcs to continue the last. (See Note 3.)

FIG. 119.—*To describe an ellipse, one diameter, A B, and an ordinate, C E, cutting the diameter in E, being given.* (Note 4.)

Bisect A B in F by a line parl. to C E,—make B G equal to a mean propnl. between A E and E B, (fig. 24)—draw a line through G parl. to C E,—draw a line through O parl. to A B, to cut the line through G in H,—from B draw a line through H, to cut the line through F in I,—make F K equal F I,—I K is the conjugate diameter to A B. (The curve may be found by Fig. 117.)

FIG. 120.—*To find the curve of an ellipse by points, when one axis, A B, and a point C in the curve, are given.*

Bisect A B in D,—from D, with radius D A, describe a circle,—from C draw a line perpr. to A B, to cut the circle in E, on the same side of A B,—join E and D,—from C draw a line parl. to A B, to cut E D in F,—from D, with F D as radius, describe a second circle,—a diameter, G H, perpr. to A B, is the second axis,—the figure may be completed by Fig. 116.

FIG. 121.—*To describe an ellipse by the use of string and pins, the axes, A B and C D, crossing in E, being given.* (See Note 5.)

From C or D, with radius A E, cut A B in points F, F,—these points are the foci of the ellipse; insert pins in points F, F, and C,—fasten a string or thread round these three pins, without attaching it to either of them,—remove the pin at C, and substitute a marking point, with which the curve of the ellipse may be described, always keeping the string extended, and moving the point through the four points A, C, B, and D.

Note 1.—Diameters are conjugate to each other, when the one is parl. to the tangents at the extremities of the other.

Note 2.—The student should observe, that this is quite equivalent to describing an ellipse in an oblong or rhomboid.

Note 3.—The curve in this figure is continuous, because the centres from which the arcs are described, are both found on one radius from the point of junction of the arcs, or in other words, both arcs have a common tangent at their junction.

Note 4.—An ordinate is a line, from a point in a diameter, to the circumference parl. to the tangent through the extremity of the diameter.

Note 5.—This is a very good method for large purposes, though ill adapted for small ones.

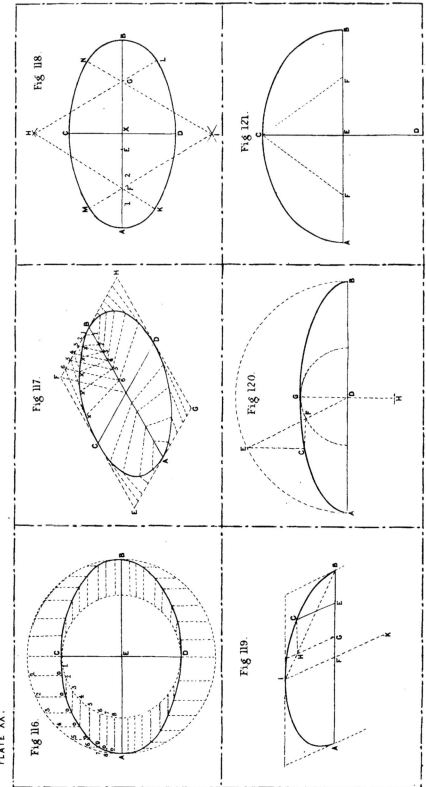

PLATE XX.

Fig 116.

Fig 117.

Fig 118.

Fig 119.

Fig 120.

Fig 121.

FIG. 122.—*To describe an ellipse, its foci, F 1 and F 2, and one point, A, in the curve, being given,*

Join points A and F 1,—draw a line from F 2 through A,—make A B equal A F 1,—bisect line B, F 2 in C,—draw a line through points F 1 and F 2, producing it both ways,—bisect line F 1, F 2, in D,—from D, with radius C B, mark points E and G,—line E G is the major axis. (The ellipse may be (fig. 120) completed by one of the previous methods.

FIG. 123.—*To describe an ellipse through any three points, not in a right line, as* 1, 2, *and* 3.

Draw line 1, 2,—bisect 1, 2 in A,—draw a line from point 3 through A,—make A B equal A 3,—lines 1 2, and 3 B are (fig. 117) conjugate diameters,—complete the ellipse.

FIG. 124.—*To find the curve of an ellipse, by the use of a paper or lath trammel, the axes being given.*

Take any strip of paper or lath, as A B C, making A B equal half the minor axis, and A C equal half the major axis,—place the lath across the diameters, so that point C, the end of the long semi-axis, coincides with the shorter diameter, and point B, the end of the shorter semi-axis, coincides with the longer diameter; the third point A, will then give one point in the curve, which may be marked on the paper; any number of points may be found in this way, by shifting the lath, and a curve drawn through them, remembering always, that the end of the long line, is on the short diameter, and the end of the shorter one, on the long diameter. (See Note.)

FIG. 125.—*To describe an ellipse about any rectangle, A B C D.*

Bisect lines A B and C D in points E and F,—draw a line through E and F, producing it both ways,—draw F B and produce it,—make B G, the part produced, equal B E,—bisect F G in H,—bisect E F in I,—from I, with radius F H, mark points K and L on E F produced,—complete the ellipse with (fig. 120) K L as one axis, and A B C and D as points in the curve.

FIG. 126.—*To describe an approximate ellipse by arcs of circles, about a given rectangle, A B C D.*

From B and C, with B C as radius, describe arcs cutting in E,—draw lines B E and C E,—bisect A B and C D in F and G,—draw F G, cutting B E and C E in H and I,—from E, with radius E B, describe arc B C,—from H and I, with radius H B, describe arcs A B and C D,—from A and D, with radius A D as radius, describe arcs cutting in K,—from K, with radius K A, describe arc A D.

FIG. 127.—*To describe an approximate ellipse, about any two equal circles given, as A and B.*

Join the centres of the circles A and B,—from A and B, with radius A B, describe arcs on either side of A B, cutting in C and D,—from C and D draw lines through A and B, to cut the circles in points E, F, G, and H,—from C and D, with radius C G, describe arcs E F and G H, continuing the arcs G E and F H.

Note.—This is a very useful method in some cases when only a small portion of the curve is required.

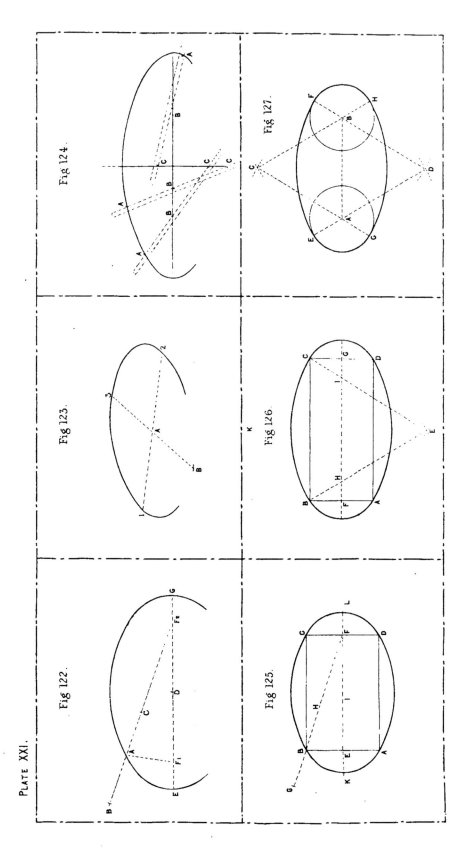

PLATE XXI.

Fig 122.

Fig 123.

Fig 124.

Fig 125.

Fig 126.

Fig 127.

FIG. 128.—*To find the centre and axes of a given ellipse.*

Draw any two chords, A B and C D, parl. to each other,—bisect A B in E, and C D in F,—draw a line through E and F, producing it to cut the ellipse in points G and H,—bisect line G H in I,—point I is the centre of the ellipse,—from I, with any radius sufficiently great, describe an arc to cut the ellipse in points 1, 2, and 3,—draw lines 1 2, and 2 3,—draw lines through I parl. to 1 2, and 2 3, to form the axes of the ellipse.

FIG. 129.—*To draw a tangent to an ellipse, at any point, Z, in the curve, and a line perpr. to the curve, at any point, Y.*

(fig. 128)
(fig. 121)

Draw the axes to the ellipse,—find the foci, F and F,—draw lines from F and F to Z, producing one as F G,—draw a line bisecting angle G Z F, and it is the tangent required.—From points F and F draw lines F H and F I through point Y,—draw a line bisecting angle H Y I, and it will be perpr. to the curve of the ellipse.

FIG. 130.—*To draw tangents to an ellipse, from a point, A, without it.*

(fig. 128)

Find the centre, B, of the ellipse,—from A draw a line through B, to cut the ellipse in C and D,—make B E a third propnl. to A B and C B,—draw any chord, F G, parl. to

(fig. 28) the diameter, C D,—bisect F G in H,—draw a diameter to the ellipse through H and B, and it will be conjugate to diameter C D,—through E draw a double ordinate, K I, parl. to H B,—from A draw tangents through points K and I.

FIG. 131.—*To describe an approximate oval by arcs of circles, its longest diameter, A B, being given.*

Trisect line A B in points 1 and 2,—through point 1 draw a line perpr. to A B,—from 1, with radius equal A 2, mark points

Note 1.—A B in this figure is a double ordinate to the parabola, and O D is an abscissa of the axis.

C and D,—bisect line 2 B in E,—from C and D draw lines through E,—from point 1, with radius 1 A, describe a semi-circle, H A I,—from C and D, with radius C I, describe arcs I F and H G,—from E, with the radius E F, describe arc G B F.

FIG. 132.—*To find the curve of an oval by points, its axes, A B and C D, intersecting in E, being given.*

This is really two semi-ellipses, and may be completed by the method used in Figure 117.

FIG. 133.—*To find the curve of a parabola by points, a base, A B, and altitude, C D, being given.*

Let line C D be perpr. to and bisecting A B in C,—at A and B erect perprs. to A B,—through D draw a line parl. to A B, to cut the perprs. in E and F,—divide C B into any number of parts, as 1, 2, 3, 4, 5, and 6,—divide B F proportionally to C B, in a, b, c, d, e, and f,—from points 1, 2, 3, &c., draw lines parl. to C D, and from each corresponding point in B F, draw a line to point D, to cut the verticals, in points o, o, o, &c., and repeat the process on the other side. (See Note 1.)

FIG. 134.—*To find a hyperbola by points, the axis, or diameter, A B, a base, B C, and an abscissa, D B, being given.*

At C erect C E; perpr. to B C, and equal to B D,—divide B C into any number of parts, and divide C E proportionally to B C,—from the points in B C draw lines to point A, and from the corresponding points in C E, draw lines to point D, cutting the lines to A in o, o, o, &c. (See Note 2.)

Note 2.—Line B O in this figure is an ordinate of the hyperbola, and to complete the curve the same work should be repeated on the opposite side of A B.

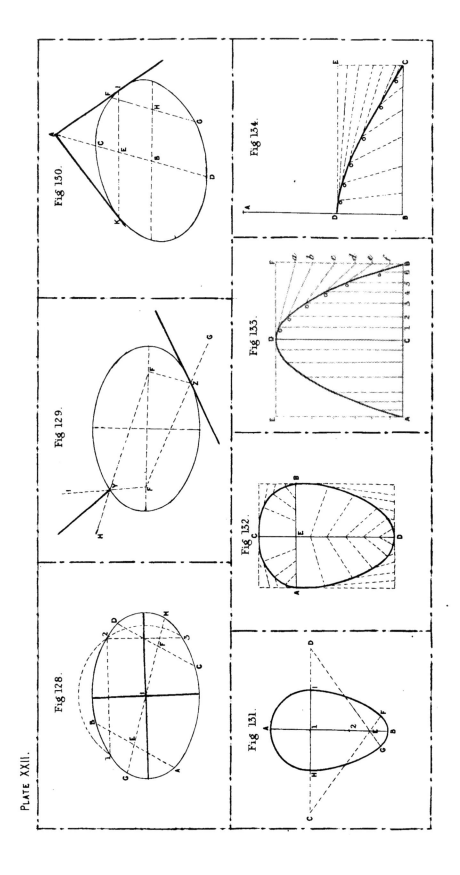

PLATE XXII.

Fig 128.

Fig 129.

Fig 150.

Fig 131.

Fig 152.

Fig 133.

Fig 134.

Fig. 135.—*To find the axis of a given parabola.*

Draw two chords parl. to each other, as A B and C D,—bisect A B in E, and C D in F,—draw a line through E and F,—draw any line, G H, perpr. to E F,—bisect G H in K,—through K draw the axis parl. to E F.

Fig. 136.—*To draw a tangent to a parabola, first, from a point in the curve, at A, and, second, from a point outside it, as B.* (See Note.)

(fig. 135) Find the axis C D, produced beyond the curve,—from A draw a line perpr. to the axis, to cut it in E,—on the axis produced mark O F, equal O E,—draw the tangent from F through A.—Second, from B draw a line parl. to the axis, to cut the curve in G,—make G H equal B G,—draw a tangent to the curve through G,—through H draw a line parl. to the tangent through G, to cut the curve in points I and K,—draw tangents from B through points I and K.

Fig. 137.—*To draw the radiating lines for the joints in a parabolic arch.*

(fig. 135) Draw the base line, A B, and the axis, C D, produced downwards,—at A or B erect a line, B E, perpr. to A B,—divide the curve of the arch into any odd number of equal parts, as may be desired,—number these parts, from the vertex down one side, as 1, 2, 3, 4, &c.,—from the divisions 1, 2, 3, &c., draw lines perpr. to E B, to cut it in a similar number of points, 1, 2, 3, 4, &c.,—draw O B, and B F, perpr. to it,—bisect B F in G,—mark on the axis O H, equal B G,—from H, mark the

divisions 1, 2, 3, 4, 5, and 6, equal to the divisions on E B,—from the points 1, 2, 3, 4, &c., on the axis, draw lines, on either side, through the corresponding numbers on the curve.

Fig. 138.—*To continue the curve of a parabola, a part including the vertex, A, being given.*

Draw the axis, A B,—draw a chord, C D, perpr. to the axis,—draw A D, and a line, D E, perpr. to it, cutting the axis in E,—draw a line through E, perpr. to the axis,—from E, with radius E D, mark this line in F and G, which are points in the curve, line F G, like line C D, being a double ordinate to the parabola,—in a similar way, the line G H, being drawn perpr. to G A, and a line drawn through point H, parl. to F G, and marked on either side of H, equal to H G, in points I and K, we have a third double ordinate; intermediate points in the curve may be found similarly, or by the Figure 133.

Fig. 139.—*To describe a parabola to touch the two sides of a given angle, A B C, and one in a given point, D.*

Draw B E, bisecting the angle A B C,—from D draw a line perpr. to B E, to cut it in F,—bisect F B in G,—point G will be the vertex, G E the axis, and a line, D F, an ordinate of the parabola, which may be continued by the same means as the preceding figure.

Note.—The perpr. to a tangent at the point of contact, is perpr. to the curve.

PLATE XXIII.

Fig 135.

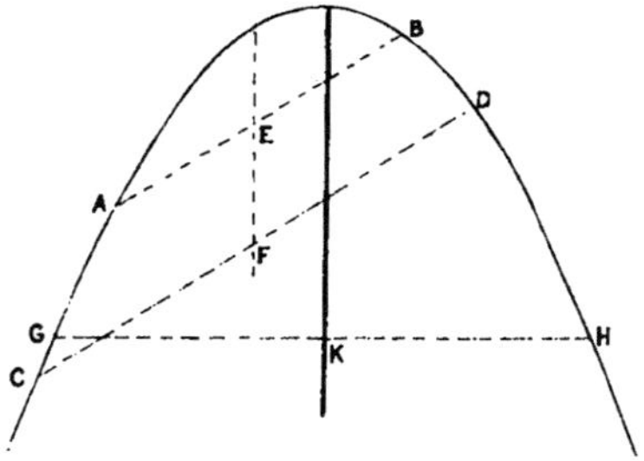

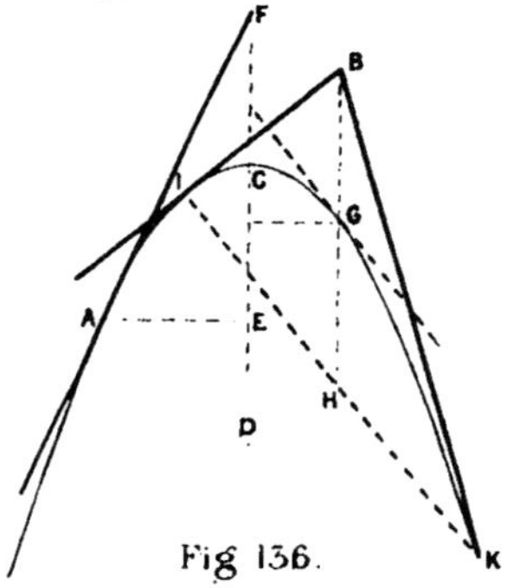

Fig 136.

Fig 137.

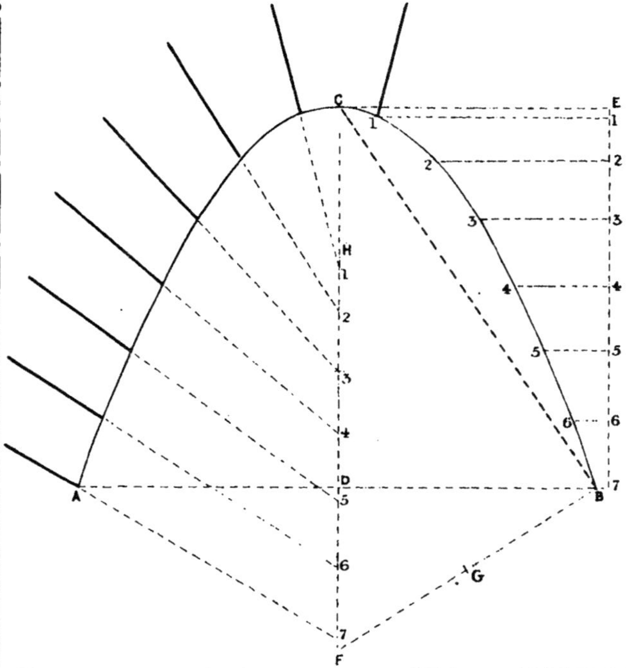

Fig 138.

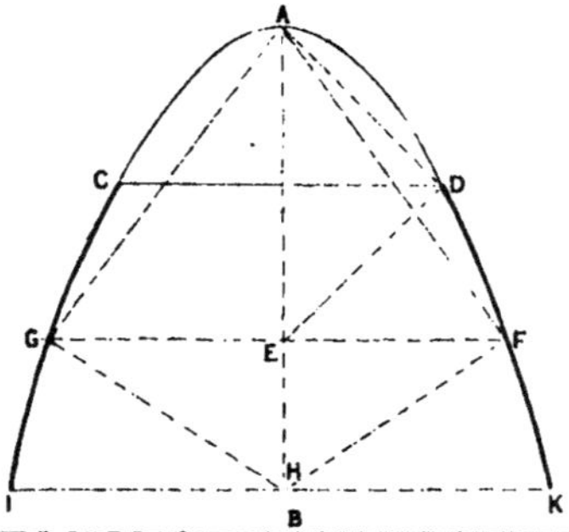

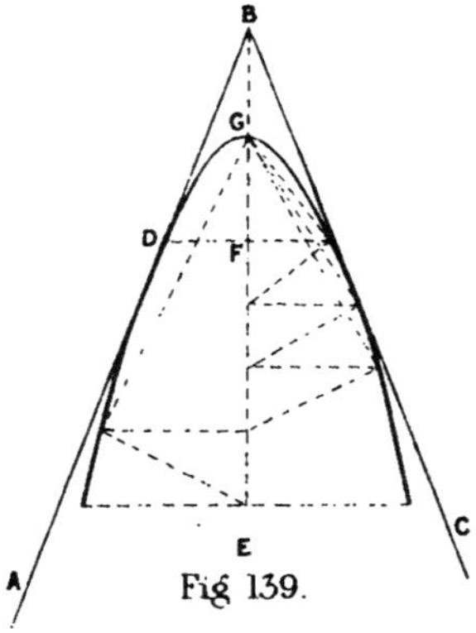

Fig 139.

FIG. 140.—*To find a curve equidistant from a given point, A, and a given line, BC.*

From A draw a perpr., AD, to BC,—bisect AD in E,—from E, on either side, set off a number of equal parts, as 1, 1, 2, 2, 3, 3, &c.,—from A, with radius A1, beyond point E from A, describe an arc,—through point 1, beyond E from BC, draw a line to cut the arc in points O and O,—similarly, from A describe a series of arcs through all the points beyond E, and cut them by parls. to BC, through all the corresponding points beyond E, from BC, in O, O, O, &c.,—through points O, O, O, draw the curve. (See Note 1, next page.)

FIG. 141.—*To find a curve equidistant from a given line, A B, and a given circle, C.*

From C, the centre of the given circle, draw a line, OD, perpr. to AB, cutting the circle in E,—bisect ED in F,—from F, on both sides, set off a series of equal points, as $a, a, b, b, c, c,$ &c., through the points beyond F, towards line AB, describe arcs from point C as centre, and through the points from F to C, draw lines parl. to AB, to cut the arcs in points O, O, O, &c.,—which are points in the curve. (See Note 2, next page.)

FIG. 142.—*To find a curve equidistant from two given unequal circles, A and B.*

Join the centres, A and B, by a line cutting the circles in C and D,—bisect OD in E,—from E, on either side, set off equal distances, and describe arcs from the centres, A and B, through the corresponding points on the opposite sides of E, cutting in points O, O, O, &c, and draw the curve as before.

FIG. 143.—*To find a curve within a given circle, A, equidistant from the circle, and from a given point within it, as B.*

From A, the centre, draw a line through B, producing it to cut the circle in O,—bisect BO in D,—from D set off a

number of points towards A, as 1, 2, 3, 4, &c.,—from A describe arcs through all these points,—from B, the given point, with radius C1, cut the arc through 1, in points O and O,—from B, with radius O2, cut the arc through 2, in points P and P,—from B, with radius O3, cut the arc through point 3, in points R and R,—similarly, with the arcs through 4, 5, 6, &c., cut them by arcs described from point B, with the radii O4, O5, O6, &c.,—through the points O, O, P, P, R, R, &c., draw the curve,—the curves on either side, will meet in a point, E, the point of bisection of line B F.

FIG. 144.—*To find a conchoid curve, its asymptote, A B, its axis, O, and its diameter, D E, being given.*

On AB mark any number of points, as 1, 2, 3, &c.,—draw a line, CF, perpr. to AB, and produce it,—from F, with radius DE, mark points G and H,—draw lines from point O through points 1, 2, 3, &c., producing them beyond AB,—from the points 1, 2, 3, &c., with radius DE, mark on these lines points O, O, O, &c., and P, P, P, &c.,—the curve drawn through the points O, O, O, is the superior conchoid, and the curve through P, P, P, is the inferior conchoid. (See Note 3, next page.)

FIG. 145.—*To find a cissoid curve, its asymptote, A B, and the generating circle, O, being given.*

Draw a diameter, DE, to the given circle, perpr. to AB,—on AB mark any number of points, as 1, 2, 3, 4, 5, 6, &c., and from all these points draw lines to point E, cutting the circle in points $a, b, c, d, e, f,$ &c.,—from point E, with radius $a, b,$ mark on line Eb point O,—from E, with radius 3 c, mark on line E 3 point O, and so continue with each line from E, marking off on it from E, the distance between the line A B, and the circle, as in each particular line. This should be repeated to the left of the line D E, and a curve drawn through all the points O, O, O, &c.

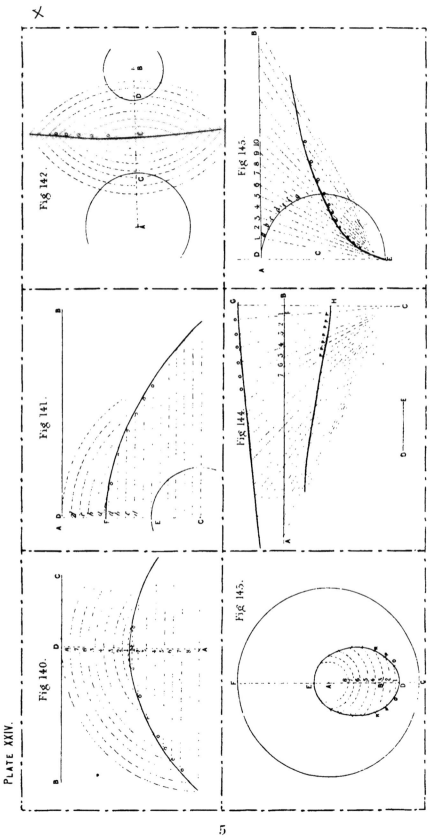

PLATE XXIV.

Fig 140.

Fig 141.

Fig 142.

Fig 143.

Fig 144.

Fig 145.

FIG. 146.—*To find the involute of a given circle, C, by points.*

Divide the circle, C, into a number of small equal parts, as 1, 2, 3, 4, 5, 6, &c.,—from all these points draw radii to the centre of the circle,—from each of the points in the circle draw lines perpr. to the radii, these are tangents to the circle,—make tangent 1 equal one division of the circle, tangent 2 equal two divisions, tangent 3 equal 3 divisions, &c., each tangent increasing one division, in points O, O, O, &c.,—through all the points O, O, O, draw the curve. This may be continued round and round the circle.

FIG. 147.—*To describe a spiral, by arcs of 60°, the increase in each arc being given, as A B.*

On A B construct a regular hexagon, A B 1 2 3 4,—produce each side of the hexagon in one direction,—from point B, with A B as radius, describe a circle cutting side 1 B produced in C,—from point 1 describe arc C D,—from point 2 describe arc D E,—from point 3 describe arc E F,—from point 4 describe arc F G, &c., using all the points of the hexagon again and again as centres.

Note 1.—The curves of Figs. 140, 141, and 142, are parabolas, the lines being the directrix, and the point, or circle, the generatrix of the parabola.

Note 2.—The student must observe, that the arcs and lines, or corresponding arcs, through the points, at equal distances from the intermediate point, as F, are the ones whose intersection he must mark. This applies to the two preceding, and the two succeeding problems.

FIG. 148.—*To describe a spiral, of a given number of revolutions, its greatest diameter, A B, being given.*

Bisect A B in C,—divide the semi-diameter, O B, into the same number of parts, plus one, that the spiral is required to have revolutions, in points D, E, &c.,—bisect O D by a perpr. in X,—from X, with radius X C, describe a circle cutting the perpr. through X in F and G,—join points O, F, D, and (fig. 148A) G, forming a square,—through X draw lines 1 3 and 2 4,* parl. to the sides of the square,—divide each of the lines 1 X, 2 X, 3 X, and 4 X, into the same number of equal parts, that the spiral is required to have revolutions, marking the points in spiral succession 5, 6, 7, 8, 9, 10, &c.,—from point 1, draw a line through point 2,—from 1, with radius 1 A, describe an arc to cut 1, 2, produced, in point H,—from point 2 produce a line through 3,—from point 2 describe arc H 1,—from 3 draw a line through 4,—from 3, with radius 3 1, describe arc 1 K,—from 4 draw a line through 5, and from 4, with radius 4 K, describe arc K L,—in a similar manner, continue with each successive point, drawing a line through the point in advance, and continuing the curve to cut the line, until at last the arc coincides with the circle in the centre, which is called the eye of the spiral.

Note 3.—To complete the curve, the same process may be extended to the right of O B, as to the left, and the curve continued to any desired extent.

Note *.—Points 1 and 2 must be marked on the extremities of these lines, nearest point A.

X

PLATE XXV.

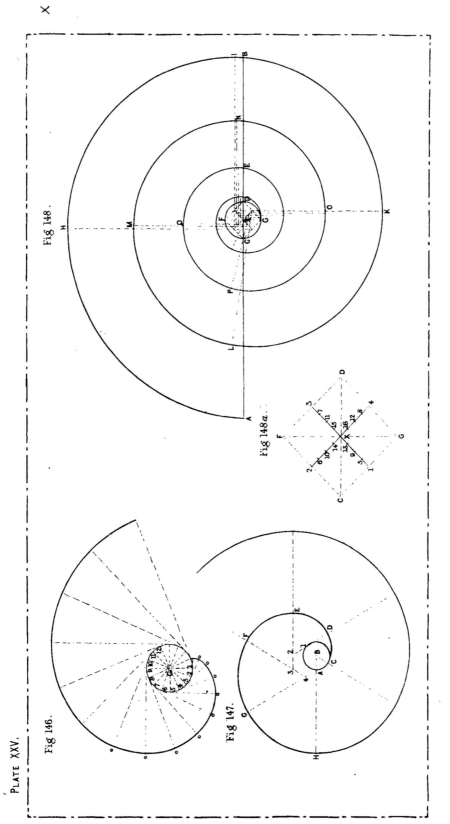

Fig 148.

Fig 148a.

Fig 146.

Fig 147.

FIG. 149.—*To describe an Archimedean spiral of any certain number of revolutions, its circumscribing circle being given.*

Divide the circle into any number of equal parts, as 16, by the radii 1, 2, 3, &c.,—divide a radius, as 1 X, into the same number of equal parts as you require revolutions in the spiral,—divide each part of the radius 1 X, into the same number of equal parts, that the whole circle is divided, as 16, —starting from the centre, on radius 2, set one of these sub-divisions,—on radius 3 set two, on radius 4 set three, on radius 5 set 4, and so on, increasing one sub-division on each radius, and afterwards, draw the curve of the spiral, through these points on the radii.

FIG. 150.—*To describe an Ionic volute, or spiral, its cathetus, A B, and the radius of the eye, A O, being given.*

Produce A B through A,—from A, with radius A O, describe a circle for the eye, having diameter O D,—bisect D A and A O in points E and F,—on E F construct a square, (fig.150A.) E G H F,—draw lines G A and H A,—trisect lines G A and H A, in points O, O,—trisect lines E A and A F in points O, O,—these points O, O, O, O, and the points F, E, G, and H, are the successive centres of the curve; they should be numbered, as in the plates, and the spiral described as in Fig. 148.

For the inner line, or web of the volute, the width, B Z, being given,—at B draw a perpr., B Y, equal B Z,—at Z draw a parl. to B Y,—draw Y O, cutting the line from Z in X,—mark F K, equal Z X,—trisect F K in r, r,—through K, r, and r, draw lines parl. to G H, to cut G A and H A, in points T, T, T, &c.,—from all the points T draw lines perpr. to E F, to cut it in other points, T, T, &c.,—all these points T are the centres of the curve of the inner spiral, which will then require to be described, in the same way as the other curve, and if they are perfectly described, both curves will coincide with the circle of the eye, in point O.

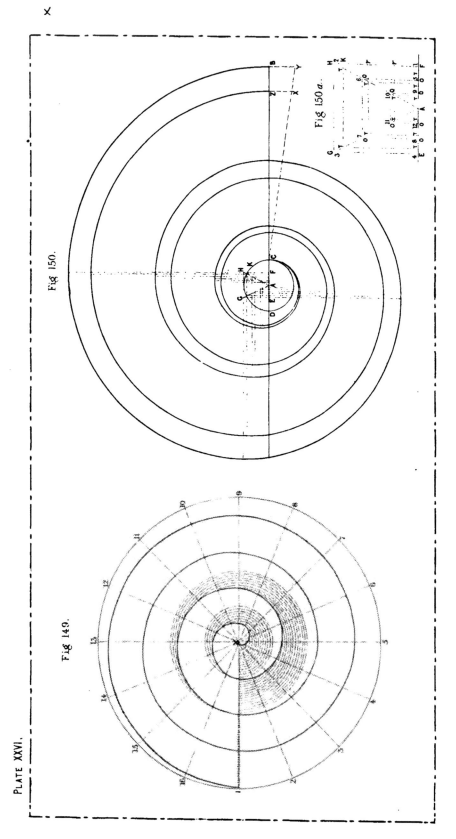

PLATE XXVI.

Fig 149.

Fig 150.

Fig 150 a.

FIG. 151.—*To find the curve traced by a point on a circle, when the circle rolls upon a straight line. (This curve is called a cycloid, the line is the directrix, and the point on the circle the generatrix.)*

Taking A B as the line, and point X, on the circle C, as the given point,—divide the circle into a number of equal parts, as a, b, c, d, e, &c., (the smaller these parts are, the more accurate will be the curve),—from centre C draw a line, C D, parl. to A B,—on C D, from C, set off a number of parts, equal to the parts of the circle, as 1, 2, 3, 4, 5, &c.,—from the points a, b, c, d, e, f, &c., draw a series of lines parl. to A B,—from point 1, with radius C X, cut the line through point a,—from 2, with radius C X, cut the line through point b,—from 3, with radius C X, cut the line through c,—continue from each succeeding point on C D, with the same radius, C X, to cut each succeeding line through points on the circle, in points O, O, O, O, &c., and draw the curve through these points.

FIG. 152.—*To find the curve traced by a point, X, within a given circle, when the circle rolls upon a right line, A B. (This is called a trochoid.)*

From the centre of the given circle, C, draw a line through X, to cut the circle in D,—from D divide the circle into a number of equal parts,—draw O E parl. to A B,—on C D, from C, set off a number of parts, equal to the divisions on the circle,—from C describe a circle through X,—draw radii from the points on the given circle, to cut the circle through X,—

draw lines from the points on circle X parl. to A B,—as in the preceding figure, from each point on O E, with radius O X, cut the lines through the divisions on circle X, beginning from 1, cut the line through the first division from X,—from 2, cut the line through the second from X, and so on with each succeeding centre and line; draw the curve through all the points so obtained.

FIG. 153.—*To find the curve traced by point X, at a distance from a given circle, but attached to it, when the circle rolls on a given line, A B. (This curve is a trochoid.)*

As in the last figures, divide the given circle into equal parts, and produce radii, to cut a circle described from C through X, mark off the parts of the given circle, on a line drawn through its centre parl. to A B,—draw lines through the points on circle X, parl. to A B,—from the divisions on line C D, with the radius C X, cut the parl. lines as before. (This figure is evidently worked to the right and left.

FIG. 154.—*To find the curve traced by a point on a circle, when the circle rolls round a second circle, outside it. (This is called an epicycloid.)*

The only variation, in this, from the preceding figures, is that the line through C, and all the lines through the divisions in the circle C, are co-centric with the directing circle, and that radiating lines, are drawn through the divisions set off on the directrix, to cut the circle through C, which points are the successive centres.

PLATE XXVII.

Fig 151.

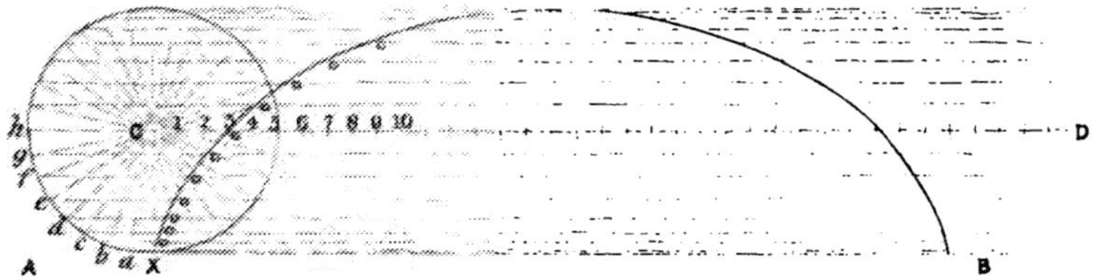

Fig 152.

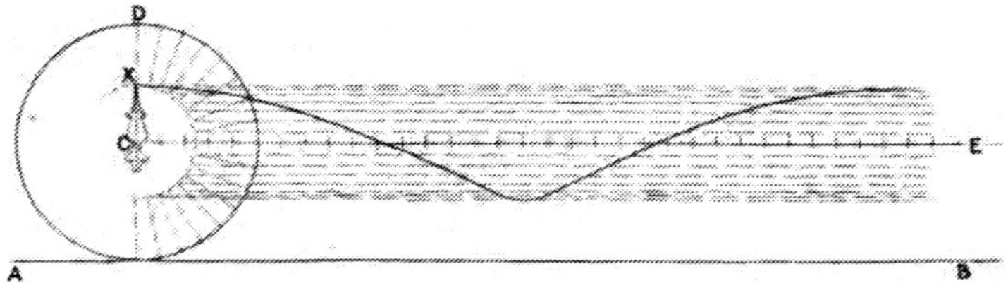

Fig 153.

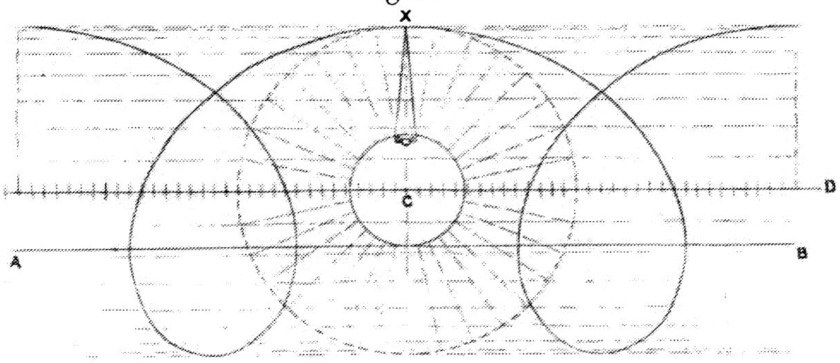

Fig 154.

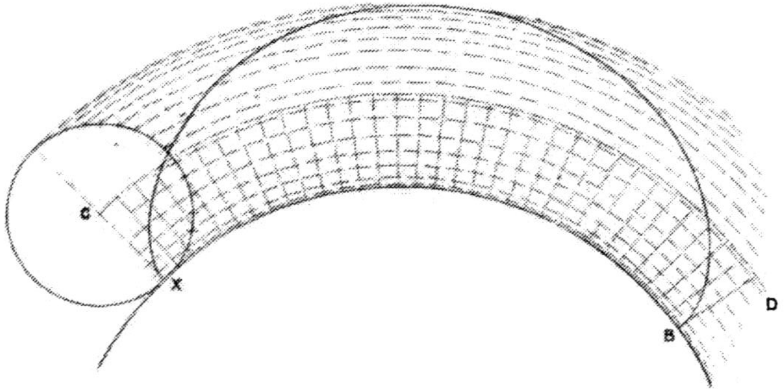

FIG. 155.—*To find the curve traced by a point, A, within a circle, when the circle rolls round, and outside of a given circle.* (*This curve is an epitrochoid.*)

This figure is worked in the same manner as Fig. 152, with the addition of the peculiarity of Fig. 154, where the divisions of the generating circle are marked on the directrix, and radiating lines drawn through the centre of the generating circle.

FIG. 156.—*To find an hypocycloid, i. e., a cycloid within the directrix.*

As in the previous figures, the generating circle is divided into a number of equal parts, which are set off, on the directing circle, and lines drawn from these points to the centre, to cut a circle, described through the centre of the generating circle; and the figure is finished as before. (See Note 1.)

FIG. 157.—*To find a curve of a trochoid, or cycloid, to go an exact number of times, in one round of the directing circle.*

It must be borne in mind, that the ratio of one circle to another, is the same as the ratios of their diameters; or, in other words, if the diameter is one-sixth of a second diameter, the perimeter, or measure of the first circle, will be one-sixth of the perimeter of the second, or any other proportion that it might be; if, therefore, it is required that either of the foregoing curves should be repeated a certain number of times in one revolution of the directrix, the generating circle must have its diameter, that proportion of the diameter of the directrix, that there are required curves. Thus, in the figure we have six hypotrochoids in one continuous line, by making the diameter of the generating circle one-sixth of the diameter of the directrix, Z Y X.

The circle, Z Y X, is divided into six equal parts, and these parts are again subdivided, into the same number of parts as the generatrix is divided into; in this case, 24 in each, and the curves are completed as before, or it may be shortened, by describing the circle through the centre of the generatrix, and dividing and subdividing it.

Note 1.—In this problem, if the generating circle has a diameter, half the diameter of the directrix, two right lines will be generated.

The student will observe, that there are some other curves, of the same kind, which have not been given, but since they may all be found by similar means, it is to be hoped, that these illustrations will be found sufficient.

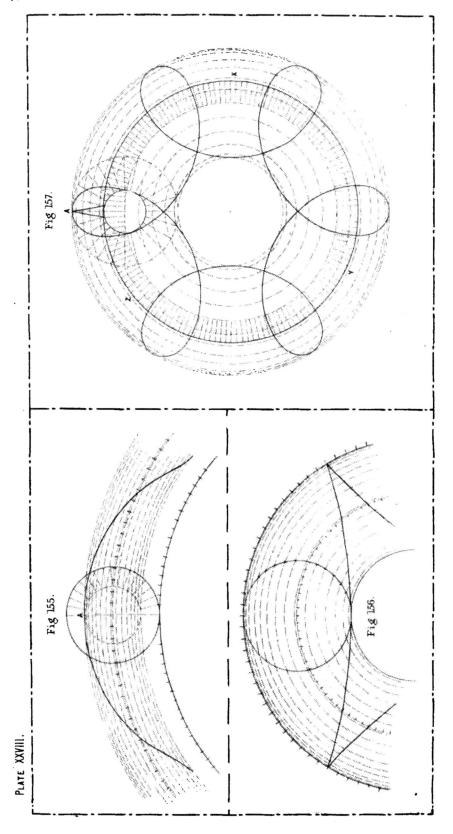

PLATE XXVIII.

Fig 155.

Fig 156.

Fig 157.

FIG. 155.—*To find the curve traced by a point, A, within a circle, when the circle rolls round, and outside of a given circle. (This curve is an epitrochoid.)*

This figure is worked in the same manner as Fig. 152, with the addition of the peculiarity of Fig. 154 where the divisions of the generating circle are marked on the directrix, and radiating lines drawn through the centre of the generating circle.

FIG. 156.—*To find an hypocycloid, i.e., a cycloid within the directrix.*

As in the previous figures, the generating circle is divided into a number of equal parts, which are set off, on the directing circle, and lines drawn from these points to the centre, to cut a circle, described through the centre of the generating circle; and the figure is finished as before. (See Note I.)

FIG. 157.—*To find a curve of a trochoid, or cycloid, to go an exact number of times, in one round of the directing circle.*

It must be borne in mind, that the ratio of one circle to another, is the same as the ratios of their diameters; or, in other words, if the diameter is one-sixth of a second diameter, the perimeter, or measure of the first circle, will be one-sixth of the perimeter of the second, or any other proportion that it might be; if, therefore, it is required that either of the foregoing curves should be repeated a certain number of times in one revolution of the directrix, the generating circle must have its diameter, that proportion of the diameter of the directrix, that there are required curves. Thus, in the figure we have six hypotrochoids in one continuous line, by making the diameter of the generating circle one-sixth of the diameter of the directrix, Z Y X.

The circle, Z Y X, is divided into six equal parts, and these parts are again subdivided, into the same number of parts as the generatrix is divided into; in this case, 24 in each, and the curves are completed as before, or it may be shortened, by describing the circle through the centre of the generatrix, and dividing and subdividing it.

Note I.—In this problem, if the generating circle has a diameter, half the diameter of the directrix, two right lines will be generated.

The student will observe, that there are some other curves, of the same kind, which have not been given, but since they may all be found by similar means, it is to be hoped, that these illustrations will be found sufficient.

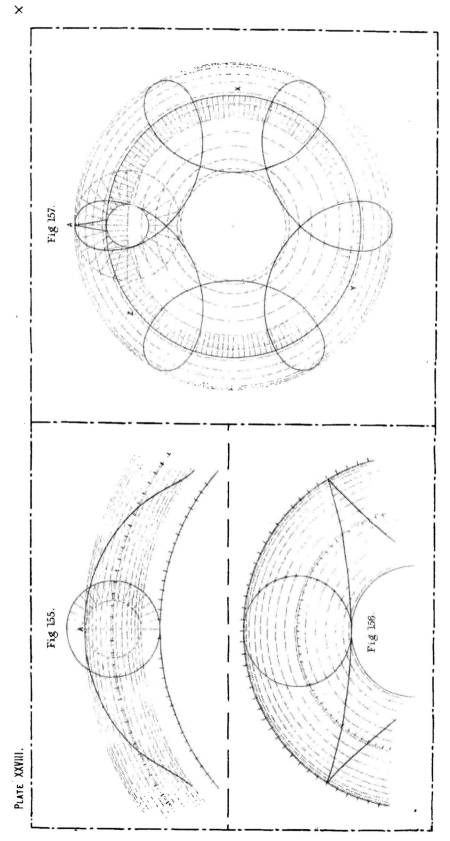

PLATE XXVIII.

Fig 155.

Fig 156.

Fig 157.

Fig. 158.—*To find the centre of any given circle.*

From any two points in the circle, as A and B, with the same radius, describe two arcs cutting in O and D,—draw C D, producing it to cut the circle in points E and F,—bisect E F in G,—G is the centre of the circle.

Fig. 159.—*To find the centre of an arc of a circle.*

Take any three points in the arc, as A, B, and O,—bisect the arcs A B and B O by perprs., producing them to meet in a point, D,—D is the centre.

Fig. 160.—*To draw a tangent to a circle, at any point on the circle, as A.*

Draw a radius, A B,—through B draw a line perpr. to A B.

Fig. 161.—*To draw a tangent to a circle, from any point without it, as A.*

Join A to the centre of the circle, B,—bisect A B in C,—from C, with C A as radius, describe an arc to cut the circle in point D or E,—draw a line from A to D or E, either will be tangential to the circle.

Fig. 162.—*To draw a tangent to an arc of a circle, from a point on it, without using its centre.*

If the point is at one extremity of the arc, as A,—join A to any opposite point, as B,—bisect A B in C, by a perpr. cutting the arc in D,—join D and A,—make angle E A D equal angle DAC,—E A is the tangent. If from a point within the arc, as Z, from Z, with any radius, mark points Y and X,—join Y X, and through Z draw a line parl. to Y X.

Fig. 163.—*To draw two tangents to a given circle, to make with each other an angle equal to a given angle, X.*

Draw a diameter, A B, and a radius, C D, making with the diameter an angle, B C D, equal the given angle, X,—from D and A draw tangents to meet in E,—angle D E A is equal angle X.

Fig. 164.—*To draw a tangent to an arc of a circle, from a point outside it, as A, without using the centre.*

From A draw a line to cut the arc in two points, as C and B,—find a mean propnl. between line A C and line A B, as A D,—from A, with radius A D, mark on the arc point E,—draw A E, and it is the tangent.

Fig. 165.—*To draw internal, and external tangents, to two given equal circles.*

Join the centres of the circles by line A B,—from A and B draw radii A C and B D, perpr. to A B,—line C D is an external tangent; for the internal, bisect A B in E, and draw a tangent to both circles through E.

(fig. 161)

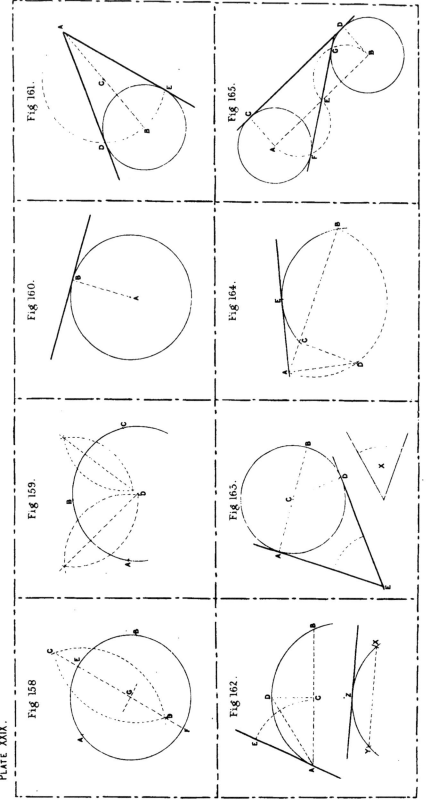

PLATE XXIX.

Fig 158.

Fig 159.

Fig 160.

Fig 161.

Fig 162.

Fig 163.

Fig 164.

Fig 165.

FIG. 166.—*To draw a tangent to two unequal circles touching each other.*

Join the centres of the circles A and B by a line cutting them in the point of contact, C,—on A B describe a semicircle —at C draw a perpr. to A B, to cut the semicircle in D,—from D, with D C as radius, describe an arc cutting the given circles in points E and F,—a line through E and F is the tangent required.

FIG. 167.—*To draw a tangent to two unequal circles cutting each other.*

Draw a line through the centres, A and B, and produce it through the smaller circle,—from A and B draw two radii, A C and B D, parl. to each other,—draw line C D, continuing it to cut A B produced in E,—on A E describe a semicircle cutting the larger circle in F,—line F E will be tangential to both circles.

FIG. 168.—*To draw an interior tangent to two unequal circles not touching each other.*

Join the centres, A and B,—on A B describe a semicircle,— make A C equal to the radii of the two circles together,—from A, with radius A C, describe an arc to cut the semicircle in D, —draw D A, cutting the given circle in E,—from B draw a line parl. to A D, to cut the second given circle in F,—draw the tangent through F and E.

FIG. 169.—*To draw an exterior tangent to two unequal circles not touching.*

Join the centres, A and B,—on A B describe a semicircle,— make A C equal to the difference between the radii of the two

Note 1.—If the longer tangent is required to touch the larger circle, the greater of the two ratios must be set up, from the larger circle, and *vice versa.*

circles,—from A, with A C as radius, cut the semicircle in D, —draw A D, producing it to cut the given circle in E,—from B draw a radius, B F, parl. to A E,—draw the tangent F E.

FIG. 170.—*To find the locus of points, from which two tangents, in a given ratio to each other, may be drawn to two given circles, A and B.* (Note 3.)

Draw a line through the centres of the circles, A and B, producing it outwards to cut the circles in points C and D,—at C and D draw lines perpr. to C D, making them, in C E and D F, in the given ratio (See Note 1),—from A, with radius A E, and from B, with radius B F, describe arcs cutting in G,—on D F make any point, H,—divide C E in I, propnlly. to F D divided in H,—from B, with radius B H, and from A, with radius A I, describe arcs cutting in K,—join G K,—bisect G K by a perpr, cutting line C D in L,—from L, with radius L G, describe a circle, which will be the locus required,—lines Z Y, Z X, from a point, Z, in this circle, are tangents in the given ratio to each other.

FIG. 171.—*To describe a circle passing through any three points, A, B, and C, not in a right line.*

Draw lines A B and B C, and bisect them by perprs, produced to meet in D,—from D, with radius D A, describe the circle. (See Note 2.)

Note 2.—This is exactly equivalent to describing a circle about any triangle.
Note 3.—A locus is a right line, circle, or curve, every point in which, and none other, satisfies a certain condition.

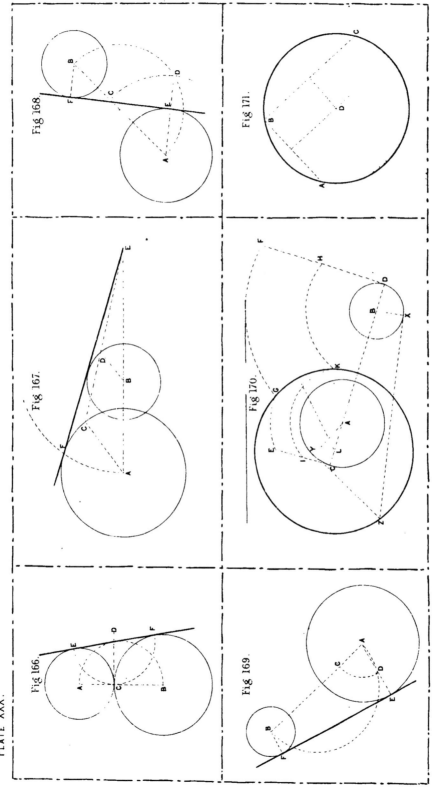

PLATE XXX.

Fig 166.

Fig 167.

Fig 168.

Fig 169.

Fig 170.

Fig 171.

FIG. 172.—*To find the arc of a circle by points, to pass through any three points, A, B, and C, not in a right line, when the centre is inaccessible.*

Draw the triangle A B C,—from A, with any radius, describe an arc cutting A B in D,—from C, with the same radius, describe an arc cutting C B in E,—from D and E, with any radius, set off on either side of D and E a number of equal measurements, as 1, 2, 3, &c.,—from A draw a line through point 1, outside D,—from C draw a line through the point 1, on the inner side of E, producing it to cut line A 1 in point F,—similarly, draw lines from A and C through points 2 and 2, to meet in G, and through as many of the other points as may be necessary, and then draw the curve through the points obtained, and points A, B, and C.

FIG. 173.—*To continue the arc, A B, of a circle, by points, when the centre is inaccessible.*

Join A and B,—from A, with any radius, set off a number of equal parts on the arc, as 1, 2, 3, &c.,—join a point, as 5, to B, the point through which the arc is to be continued,—from point 1, with radius A B, and from point 6, with radius 5 B, describe arcs cutting in C, beyond B,—from points 2 and 7, with the same radii as before, describe arcs cutting in D,—points C and D are points in the continuation of the arc, and any number may be found in a similar manner by describing intersecting arcs from the succeeding points.

FIG. 174.—*From any point, A, without a given circle, to draw a line to cut the circle in a chord of a given length, as B C.*

From the centre, X, of the given circle, describe a concentric arc, to pass through A,—from any point in the circle mark off arc D E, equal B C,—produce D E to cut the arc through A in F, from A, with radius E F, cut the circle in G,—draw A G, making chord H G the required length.

FIG. 175.—*To draw a line from a point, A, without a given circle, to a point on the circle, so that the chord may be in a given ratio to the whole line, as Z Y : Y X.*

Find a mean between the two ratios together, and the lesser one, as X W,—draw line A B, a tangent to the given circle,—find a fourth propnl. to X W, A B, and X Z, in line X V,—from A, with X V as radius, cut the circle in O,—draw line A C.

FIG. 176.—*To draw a chord through a point, A, within a given circle, such that the chord shall be divided in point A, in two segments, in a given ratio, as Z Y : X W.*

Draw a diameter, B C, through point A,—make A D a fourth propnl. to the given ratios, and the line A B,—find a mean, A E, between A D and A C,—from A, with A E as radius, cut the circle in F,—draw the line from F through A, to cut the circle in G.

FIG. 177.—*To find a point within a given circle, through which all chords that may be drawn, will have their tangents meeting in a given line.*

From the centre of the circle A, draw A B, perpr. to X Y, the given line,—draw B C and B D, tangents to the circle,—draw C D, cutting A B in E,—point E is the point required.

FIG. 178.—*To describe a circle to cut a given circle in two equal parts, and to pass through two given points, A and B, without it.*

Draw a line through the centre of the given circle, C, from point A,—find C D a third propnl. to A C, and the radius of the circle,—from C, on A C produced, mark O E equal O D,—join points E and B,—bisect A E and E B by perprs., to intersect in F,—from F, with F A as radius, describe the required circle, cutting the given circle in points G and H.

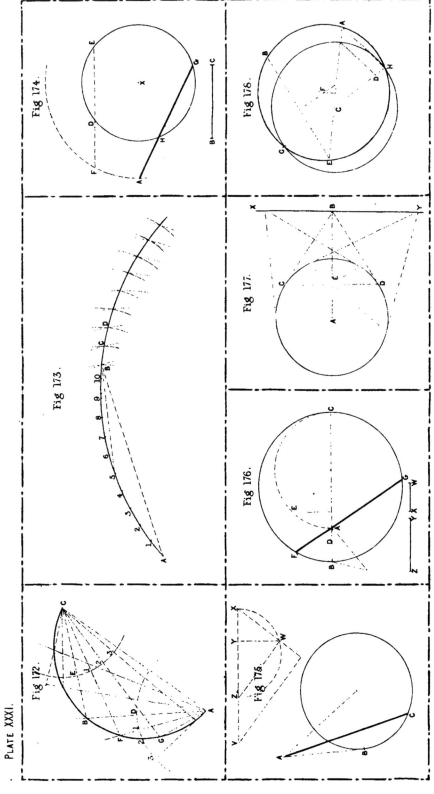

PLATE XXXI.

Fig 172.

Fig 173.

Fig 174.

Fig 175.

Fig 176.

Fig 177.

Fig 178.

Fig. 179.—*From any two points, A and B, on the given circle, to draw lines to a third point on the circle, in a given ratio to each other.*

Draw the chord A B,—bisect the arc A B, on either side of the line A B, in C,—divide line A B, in the given ratio, in D, —draw a line from C through D, to cut the circle in E,—draw lines E A and E B.

Fig. 180.—*Through one of the points of intersection of two given circles, to draw a line that is cut by the circles, in segments, in a given ratio to each other, as X : Y.*

Join the centres of the given circles, A and B,—divide A B, in the required ratio, in C,—join point C to the point of intersection of the circles, D,—through D draw a line perpr. to C D, to be cut by the circles in points E and F, in the given ratio.

Fig. 181.—*To describe a circle of a given radius, A B, to touch two convergent lines, C D and E F.*

Draw two lines, parl. to C D and E F,—at a distance from them equal line A B,—to intersect in G, with A B as radius, describe the circle.

Fig. 182.—*To describe a circle through a given point, A, to touch two convergent lines, B C and D E, one on each side of A.*

(fig.12)
(fig.13) Draw line F G, bisecting the angle formed by lines B C and D E,—through A draw a line converging to the same point as lines B C and D E,—from any point, I, in F G, describe a circle touching the lines O B and D E, and cutting the line

through A in H,—join H to the centre of the circle, I,—draw A K parl. to H I,—from K, with K A as radius, describe the circle.

Fig. 183.—*To describe three circles of given radii, as lines 1, 2, and 3, to touch each other.*

Make line A B equal lines 1 and 2 joined in C,—from A and B, with radii A C and B C, describe two of the circles,— from C, with radius equal line 3, mark on either side of C points D and E,—from A and B, with radii A E and B D, describe arcs cutting in F,—from point F, with radius line 3, describe a circle, which will touch the other two.

Fig. 184.—*To describe a circle of a given radius, A B, to touch two given circles.*

Join the centres of the circles, C and D, by a line cutting the circles in E and F,—add the line A B to each of the radii of the circles, in lines F H and E G,—from C and D, with radii C G and D H, describe arcs crossing in I,—from I, with radius A B, describe the circle.

Fig. 185.—*To describe a series of circles, to touch each other, and two convergent lines, A B and C D.*

(fig.12) Draw line E F, bisecting the angle formed by the given lines, —from any point, as G, in E F, draw a line, G H, perpr. to A B,—from G, with G H as radius, describe a circle cutting E F in I,—from I draw I K, perpr. to E F,—from K, with radius I K, mark point L on A B,—from L draw L M, perpr. to A B —from M, with M I as radius, describe a circle, cutting E F in N, and similarly proceed with the other circles.

PLATE XXXII.

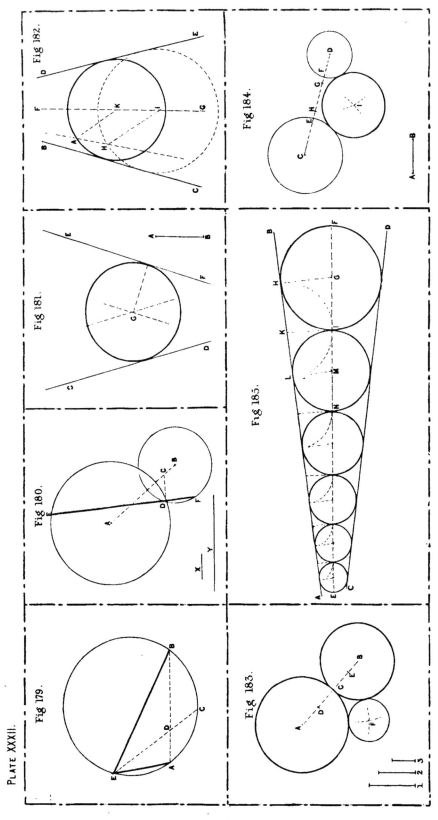

Fig 179.

Fig 180.

Fig 181.

Fig 182.

Fig 183.

Fig 185.

Fig 184.

FIG. 186.—*To describe a circle, to touch any three lines making angles with each other.*

Let the lines be produced to meet in points A and B,—draw lines bisecting the angles formed by the given lines at A and B, and produce them to intersect in C,—from C draw C D, perpr. to one of the given lines,—from C, with C D as radius, describe the circle.

FIG. 187.—*To describe a circle of a given radius, A B, to touch a given line, C D, and to pass through a given point, E.*

At a distance from C D, equal line A B, draw a parl. to C D,—from E, with radius A B, describe an arc to cut the parl. in F,—from F, with radius A B, describe the circle.

FIG. 188.—*To describe a circle, to touch a given line, A B, in a given point, C, and to pass through a given point, D.*

From point C erect a perpr. to A B,—join C and D,—bisect C D by a perpr., to cut the perpr. from C in E,—from E, with radius E C, describe the circle through D.

FIG. 189.—*To describe a circle, to pass through two given points, A and B, equidistant from a given line, C D, and to touch the given line.* (See Note 1.)

Join points A and B,—bisect A B in E, by a perpr. produced to cut line C D in F,—join A and F,—bisect A F by a perpr. to cut E F in G,—from G, with radius G A or G F, describe the circle.

Note 1.—In these figures it is evident that the two points must be on the same side of the line.
Note 2.—In both these cases there is, of course, a minimum for the line A B, less than which it must not be given.
Note 3.—The term "mean" is an abbreviated expression for mean proportional.
Note 4.—It must always be observed that, in the following problems, as in these two, the given circle is designated by a single letter, as X, and this is always taken as marking the centre of the circle.

FIG. 190.—*To describe a circle, to touch a given line, A B, and to pass through any two points, C and D, that are not equidistant from A B.* (See Note 1.)

Join D and C, producing the line to cut A B in E,—find the mean, E F, between E C and E D,—from E mark off E G or E H, equal E F,—bisect C D by a perpr.,—from G or H erect a perpr. to A B, to cut the perpr. bisecting C D in I or K,—from I or K describe a circle through points C and D, touching line A B. (See Note 3.)

FIG. 191.—*To describe a circle to pass through a given point, B, and to touch a given circle in a point, A.*

Join A and B,—bisect A B by a perpr., C D,—from A draw a line through the centre of the circle, producing it to cut C D in E,—from E, with radius E A, describe the circle through B.

FIG. 192.—*To describe a circle of a given radius, A B, to touch a given line C D, and a given circle, X.* (See Note 4.)

Add the radius of the circle X to the line A B in B E,—from X, the centre of the circle, with A E as radius, describe an arc,—at a distance from C D equal A B, draw a parl. to C D, cutting the arc in F,—from F, with A B as radius, describe the circle.

FIG. 193.—*To describe a circle of a given radius, A B, touching a given line, C D, and to touch and enclose a given circle, X.* (See Note 2.)

Deduct the radius of the given circle, X, from line A B in A E,—from X, with radius B E, describe an arc,—at a distance from C D equal A B draw a parl. to C D, to cut the arc in F,—from F, with A B as radius, describe the circle.

PLATE XXXIII.

Fig 186.

Fig 187.

Fig 188.

Fig 189.

Fig 190.

Fig 191.

Fig 192.

Fig 193.

FIG. 194.—*To describe a circle to touch a given line, AB, in a point, C, and a given circle, X,*

Through C draw a perpr. to AB,—make CD equal the radius of circle X,—join D and X,—bisect DX by a perpr. to cut DC produced in E,—from E, with EC as radius, describe the circle touching circle X.

FIG. 195.—*To describe a circle, to touch a given line, AB, in a point, C, and to touch and enclose a circle, X.*

At C erect a perpr. to AB, in CD,—make CE equal the radius of circle X,—join E and X,—bisect EX by a perpr. to cut CD in F,—from F, with radius FC, describe the circle.

FIG. 196.—*To describe a circle, touching a given line, AB, and a given circle, X, in any given point, C.*

From C draw CD, perpr. to AB,—draw a line from C through X,—bisect angle DCX, by a line produced to cut AB in E,—from E draw a line parl. to CD, to cut CX produced in F,—from F, with FE as radius, describe the circle.

FIG. 197.—*To describe a circle, to touch two given lines parallel to each other, and to touch a given circle, X, which is, in part or completely, between the parallels.*

Draw AB, perpr. to and joining the parls.,—bisect AB in C, by a perpr.,—from C, on CA produced, mark CD, equal the radius of the circle, X,—from the centre, X, with BD as radius, cut the line through C in E,—from E, with BC as radius, describe the circle.

FIG. 198.—*To describe a circle, about and touching a given circle, X, and touching two parallels, between which the given circle is described.*

Draw AB, perpr. to and joining the parls.,—bisect AB in C by a perpr., CD,—make CE equal the radius of circle X,—from X, with radius EB, cut line CD in F,—from F, with radius CB, describe the circle.

FIG. 199.—*To describe two circles, of given radii, as AB and CD, within a given circle, touching each other, and the given circle, X.*

Draw a diameter, EF, through X,—make EG equal AB, and FH equal CD,—from G, with radius GE, describe a circle cutting EF in I,—make IK equal CD,—from G, with radius GK, and from X, with radius XH, describe arcs cutting in L,—from L, with radius CD, describe the circle.

FIG. 200.—*To describe a circle, touching two given circles, Z and Y, and one of them in a given point, A.*

Draw a radius, ZA, producing it outwards,—make AB equal the radius of the circle, Y,—join B and Y,—bisect BY by a perpr., to cut ZA produced in C,—from C, with CA as radius, describe the circle.

FIG. 201.—*To describe a circle, touching one given circle, Z, in a given point, A, and enclosing a second circle, Y, touching it.*

Draw a radius, ZA, producing it,—make AB equal the radius of the circle, Y,—join centre Y to point B,—bisect BY by a perpr., to cut ZB produced in C,—from C, with radius CA, describe the circle.

PLATE XXXIV.

Fig 194.

Fig 195.

Fig 196.

Fig 197.

Fig 198.

Fig 199.

Fig 200.

Fig 201.

FIG. 202.—*To describe a circle, touching and enclosing two given circles, Z and Y, and touching one in a given point, A.*

Draw radius A Z, producing it,—draw radius Y B, parl. to A Z,—draw A B, producing it to cut the circle, Y, in C,—bisect A C by a perpr., to cut A Z produced in D,—from D, with D A as radius, describe the circle.

FIG. 203.—*To describe a circle of a given radius, as A B, to touch two given circles, Z and Y, and to enclose one of them.*

Make A C equal to the radius of the circle, Z, to be enclosed,—produce A B, making B D, the production, equal the radius of the circle, Y,—from Z, with radius C B, describe an arc, and from Y, with radius A D, describe an arc cutting the first arc in E,—draw a line from E through Z, to cut the circle in F,—from E, with radius E F, describe the circle.

FIG. 204.—*To describe a circle of a given radius, A B, to touch and enclose two given circles, Z and Y.*

Deduct the radius of circle Z, from line A B in A C,—with the remainder, C B, describe an arc from the centre, Z,—deduct the radius of circle Y from A B, in A D,—with the remainder, D B, as radius, describe an arc from centre Y, cutting the first arc in E,—from E, with radius A B, describe the circle.

FIG. 205.—*To describe a circle, to pass through two given points, A and B, and to touch a given circle, Z; when the points are equidistant from the circle.*

Join points A and B,—bisect A B in C,—from C draw a line through the centre, Z, producing it to cut the circle in points D and E,—join D and B,—bisect D B by a perpr. to cut the line C E in F,—from F, with radius F A, describe a circle through points A, B and D.

If E and B are joined, and E B bisected by a perpr. to cut C E in G, G will be the centre of a circle that would pass through points A and B, and enclose the given circle.

FIG. 206.—*To describe a circle, to pass through two given points, A and B, and to touch a given circle, Z, when the two points are in a line passing through the centre of the circle.*

Join A and B,—bisect A B by a perpr., C D,—from any convenient point in C D describe an arc, passing through A and B, and cutting the given circle in points E and F,—draw a line through E and F, to cut line A B produced in G,—from G draw a tangent, G H, to the given circle,—from Z draw a line through point H, to cut C D in point I,—from I, with radius I A or I H, describe the circle.

FIG. 207.—*To describe a circle, to pass through two given points, A and B, and to touch a given circle, Z, when the points are not in a line, through the centre, Z, or equidistant from the circle.*

This is identical with Fig. 206, in construction, with the addition that we may have a second tangent, G K, and a circle through points A B and K, which will enclose the circle, Z.

FIG. 208.—*To describe a circle, through two points, A and B, within a given circle, Z, and touching the given circle.*

This, again, is identical in description with the two previous figures, unless the points are equidistant from the circle.

FIG. 209.—*To describe a circle within a given circle, Z, to touch it, and any two lines crossing it, as A B and C D.*

Bisect the angle formed by the lines A B and C D, by line E F,—at a distance from C D equal the radius of the circle, Z, draw G H, parl. to C D,—from Z, the centre, draw a perpr. to E F, to cut it in I, and line G H in K,—make I L equal I Z,—find a mean, K M, between K L and K Z,—on either side of K mark point O, or P, with radius K M,—from O or P draw a perpr. to G H, to cut E F in Q or R,—either Q or R may be the centre of the circle required.

PLATE XXXV.

Fig 202.

Fig 203.

Fig 204.

Fig 205.

Fig 206.

Fig 207.

Fig 208.

Fig 209.

FIG. 210.—*To describe a circle to touch two given lines, A B and C D, and a given circle, Z, which is entirely, or in part, between the given lines:*

This is identical in description with Fig. 209, except that line G H must be outside the given line.

FIG. 211.—*To describe a circle, to pass through a given point, P, to touch a given circle, C, and to touch a given line, A B; when the point is outside the circle.*

Through centre C draw a diameter, D E, perpr. to A B, and produce it to cut A B in F,—draw a line from D through P,—find D Q, a fourth propnl. to lines D P, D F, and D E,—mark D R on D P, produced equal D Q,—produce D R to cut A B in S,—find S T, a mean between S R and S P,—make S V equal S T,—from V draw a perpr. to A B,—bisect P R by a perpr., to cut the perpr. from V in X,—X is the centre of the circle required.

FIG. 212.—*To describe a circle, to pass through a given point, P, to touch a given line, A B, and to touch and enclose a given circle, C.*

Draw a diameter, D E, perpr. to A B, produced to cut A B in F,—draw a line from P through E,—make E Q on P E produced, equal the fourth propnl. to P E, E F, and D E (fig. 29),—bisect P Q by a perpr. to cut D E in R,—R is the centre of the required circle.

FIG. 213.—*To describe a circle, to pass through a given point, P, to touch a given circle, C, and a given line, A B, when the point, P, and the line are within the circle.*

Draw diameter D E, perpr. to A B, cutting it in F,—from E draw a line through P,—make E Q a fourth propnl. to P E, E F, and E D (fig. 29),—describe a circle on either side, through points P and Q (fig. 190), touching the line A B, and it will also touch the circle C.

FIG. 214.—*To describe a circle, passing through a given point, A, and touching two given circles, B and C.*

Draw a line through the centres B and C, and produce it,—draw an external tangent (fig. 167), to touch the circles in points D and E, and to cut B C produced in F,—join points A and F,—make F G a fourth propnl. to F A, F E, and F D,—a circle described through points A and G, and touching one circle by problem 306, will also touch the second circle. (See Note 1.)

FIG. 215.—*To describe a circle to pass through point A, within two given circles, B and C, and to touch the given circles.*

Describe (fig. 142), a curve equidistant from point A and the circle B,—describe a second curve, equidistant from point A and the circle C, cutting the first curve in points D and E,—points D and E are the centres of circles that would pass through point A, and touch the given circles,—the radii may, of course, be found by drawing a line from either centre of the given circles, through D or E, to the circle.

FIG. 216.—*To describe a circle to touch three given unequal circles, A, B and C.*

Find a curve equidistant from the circles A and B (fig. 142),—find a second curve, equidistant from the circles B and C, cutting the first curve in point D,—D is the centre, and the distance to any one of them is the radius of a circle to touch the three circles.

FIG. 217.—*To describe a circle about three given circles, Z, Y, and X, and touching them.*

Join the centres Z, Y, and X, and produce the lines to cut the circles in points A, B, C, and D,—bisect A B in E, and C D in F,—from E and F, on both sides of each, set off a number of equal parts, as 1, 1, 2, 2, 3, 3, &c.,—through these points describe arcs cocentric with the circles, to cut each other in the points P, P, P, P, &c.,—through each series of points P, draw curves crossing in point G,—point G is the centre of the required circle, and the radius may be found by drawing a line through the centre of one of the circles, from G, as G H.

Note 1.—Of this problem there are several cases, as in some of the previous ones; they may be worked in a similar manner to this figure; at the same time the student may observe, that the means used in the three following figures, are, though scarcely mathematically admissible, much more practically accurate and simple, and that they may be applied to this, and several of the preceding problems, with advantage.

PLATE XXXVI.

Fig 210.

Fig 211.

Fig 212.

Fig 213.

Fig 214.

Fig 215.

Fig 216.

Fig 217.

FIG. 218.—*Within or about a given triangle, ZYX, to describe a similar triangle, one side, S, or S¹, being given.*

Bisect two of the angles of the triangle by lines to meet in A, and produced outwards,—on a side of the triangle, as YX, or YX produced, mark the given side from X, as XO or XB,—from points B and O draw lines parl. to XA, to cut AY and AY produced, in D and E,—from D and E draw lines parl. to the sides of the triangle,—the points F and G, where two of these lines cut the line AX, give the starting points for the sides parl. to the remaining side of the triangle.

FIG. 219.—*Within a given triangle, ZYX, to inscribe a triangle, similar to a given triangle, W.*

On a side of the given triangle, as ZY, outside of it, construct a triangle, ZAY, similar to the triangle W, by making two angles equal two of the angles of triangle W,—draw line AX, cutting ZY in point B,—from B draw lines BO and BD, parl. to AZ and AY,—join O and D.

FIG. 220.—*To inscribe a square in a given triangle, ZYX.*

From an angle, as Y, draw a line YA, perpr. to the opposite side,—from Y draw a line YB, perpr. to and equal to YA,—from B draw a line to the opposite angle of the triangle, Z,—cutting YX in O,—from O draw line OD, perpr. to ZX,—from O draw a line OE, parl. to ZX,—from E draw a line EF, perpr. to ZX.

FIG. 221.—*To inscribe a rectangle, within a given triangle, ZYX, one side, WV, being given.*

From an angle, as Z, make ZA, equal line WV,—draw AB parl. to ZY,—from B draw BO, parl. to ZX,—from B and O draw lines BD and OE, perpr. to ZX.

FIG. 222.—*To inscribe a segment of a given angle, within a given isosceles triangle, ZYX.* (See Note 1.)

Bisect ZY in A,—join A and X,—bisect the given angle,—at A construct, with AX, an angle equal half the given angle, cutting the side XY in B,—bisect AB by a perpr., cutting AX in O,—from O, with radius OA, describe the arc BAD,—draw BD parl. to ZY.

FIG. 223.—*To inscribe a circle in a given sector, ZYX.*

Draw line XA, bisecting angle ZXY,—at A draw a line perpr. to XA, to cut XY produced in B,—bisect angle ABX by a line to cut AX in O,—from O, with radius OA, describe the circle.

By the same method a circle is inscribed in any triangle, OA being always made perpr. to a side.

FIG. 224.—*To inscribe a semicircle in any sector, as ZYX.*

Bisect the sector by line XA,—draw BA perpr. to AX,—bisect angle XAB by a line to cut ZX in O,—draw OD perpr. to XA, cutting XA in E,—from E, with radius EO, describe the semicircle.

FIG. 225.—*To inscribe a square in a given sector, ZYX.*

Draw the chord of the arc, ZY,—make YA perpr. to ZY, and equal to ZY,—draw AX, to cut the arc ZY in B,—draw BO, perpr. to ZY,—draw lines BD and OE, parl. to ZY,—join D and E.

Note 1.—This implies the phrase, "such a segment of a circle as would contain a given angle."

Note 2.—Inscribed figures are generally supposed to be the largest possible figures of their kind, so that they may touch all the sides of the figure in which they are inscribed, or as many as it is possible.

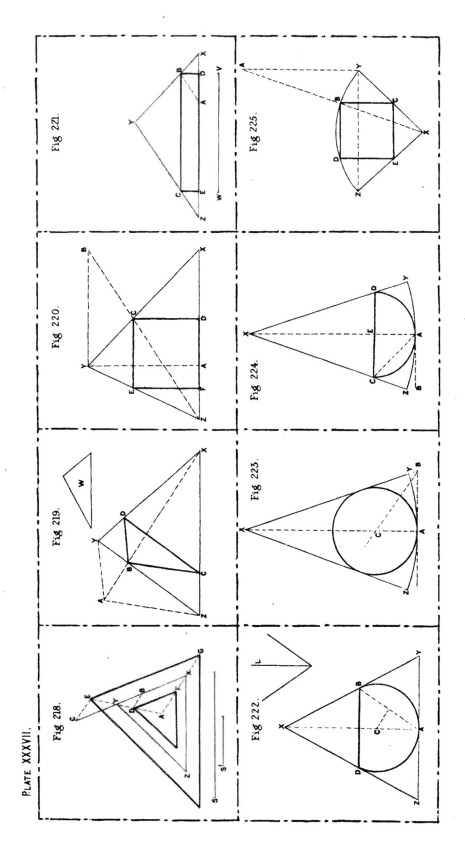

PLATE XXXVII.

Fig 218.

Fig 219.

Fig 220.

Fig 221.

Fig 222.

Fig 223.

Fig 224.

Fig 225.

PART II.—FIFTH SECTION.—FIGURES IN RELATION TO FIGURES.—PLATE XXXVIII.

FIG. 226.—*To inscribe a square in a given segment, ZYXW.* (See Note 1.)

Bisect the chord, ZW, in A,—at another point in ZW, as B, erect a perpr. to ZW,—make BC equal twice AB,—draw AC, producing it to cut the arc in D,—draw DE parl. to ZW,—draw DF and EG, perpr. to ZW.

FIG. 227.—*To inscribe a hexagon within an eqtrl. triangle, ZYX, three sides of the hexagon being coincident with the triangle.*

Trisect the sides of the triangle in points 1 2, 3 4, 5 and 6, —join points 2 3, 4 5, and 6 1.

FIG. 228.—*To inscribe a hexagon within an eqtrl. triangle, ZYX, the hexagon only touching the triangle.*

Draw lines ZA, YB, and XC, bisecting the angles and sides of the triangle, and crossing in D,—from D, with radius DA, describe a circle to cut these lines in points 1, 2, and 3, —join points 1, A, 2, B, 3, and C.

FIG. 229.—*To inscribe a duodecagon in a given eqtrl. triangle, ZYX.*

Draw ZA, YB, and XC, as before, crossing in D,—through D draw lines parl. to the sides of the triangle,—from D, with radius DA, cut all these lines in points 1, 2, and 3, &c., which when joined will give the duodecagon.

FIG. 230.—*To inscribe a duodecagon in a given eqtrl. triangle, three sides of the polygon coinciding with the sides of the triangle.*

Find points A, B, and D, as before,—draw DE, parl. to ZY,—draw DF,—from D, with radius DF, describe a circle,—from F, with radius DF, mark the circle in six equal parts, in points 1, 2, 3, 4, 5, and 6,—bisect each of these arcs in points 7, 8, 9, 10, 11, and 12,—join all these points.

FIG. 231.—*To describe a circle within or about a given square, ZYXW.*

Draw the diagonals crossing in A,—draw AB, perpr. to XW,—from A, with radius AB, and from A, with radius AZ, describe the circles.

FIG. 232.—*To inscribe an eqtrl. triangle in a square, ZYXW.*

From W, with any radius, describe a quadrant, AB,—from A and B, with the same radius, trisect the arc in C and D,—bisect arcs AC and DB, in points E and F,—from W draw lines through E and F, to cut YZ and YX in G and H,—join points G and H.

FIG. 233.—*To inscribe an isosceles triangle in a square, ZYXW, one side, AB, being given.* (See Note 2.)

From an angle of the square, as Z, with the given side, AB, as radius, describe an arc cutting two sides of the square in points C and D,—join points C and D, and point Z.

Note 1.—The given segment must not be more than three-quarters of the entire circle.

Note 2.—The given side must be less than the diagonal of the square.

PLATE XXXVIII.

Fig 226.

Fig 227.

Fig 228.

Fig 229.

Fig 230.

Fig 231.

Fig 252.

Fig 233.

Fig. 234.—*To inscribe a square in a given square,* Z Y X W, *one diagonal,* A B, *being given.* (See Note 1.)

Draw the diagonals crossing in C,—bisect A B in D,—from C, with radius D A, describe arcs to cut the sides of the given square in points 1, 2, 3, and 4,—join the points 1, 2, 3, and 4.

Fig. 235.—*To inscribe an isosceles triangle in a given square,* Z Y X W, *the base,* A B, *of the triangle being given.* (See Note 2.)

Draw a diagonal, Y W,—bisect line A B in C,—from W mark W D, equal C B,—from D, with radius C B, cut the sides of the square in points E and F,—draw E F, and join E and F to point Y.

Fig. 236.—*To construct an eqtrl. triangle about a given square,* Z Y X W.

Produce the base, Z W, both ways,—from Y and X, with Y X as radius, describe arcs cutting in A,—from A draw lines through Y and X, to cut Z W produced in B and C.

Fig. 237.—*To construct a triangle similar to a given triangle,* A, *about a given square,* Z Y X W.

On Y X construct a triangle similar to triangle A in Y B X, produce lines B Y and B X to meet line Z W produced both ways in points C and D.

Fig. 238.—*To inscribe a square in a given rhombus,* Z Y X W.

Draw the diagonals of the rhombus crossing in point A,— draw lines bisecting the angles formed by the diagonals at point A, cutting the sides of the rhombus in points B, C, D, and E,—join the points B, C, D, and E.

Fig. 239.—*To inscribe a square in a given trapezion,* Z Y X W.

Draw the shorter diagonal, Y W,—at Y draw Y A, perpr. to, and equal to Y W,—draw a line from A to the farthest angle, Z, of the trapezion, cutting side Y X in B,—from B draw B C, perpr. to Y W,—from B and C draw lines parl to Y W, to cut the sides X W and W Z in points D and E,—join D and E.

Fig. 240.—*To inscribe a square in a pentagon,* Z Y X W V.

Draw a diagonal, Z W,—at Z draw a line, Z A, perpr. to, and equal to Z W,—from A draw a line to the point V, cutting line Z Y in B,—from B draw a line perpr. to Z W, to cut V Z in C,—from B and C draw lines parl. to Z W, to cut X W and W V in points D and E,—join points D and E.

Fig. 241.—*To inscribe a square in a given hexagon,* Z Y X W V U.

Draw a diagonal, Z W,—bisect Z W by a perpr. in A,—draw lines bisecting the angles formed at A, and produce them to cut the sides of the hexagon in points B, C, D, and E,—join these four points.

Note 1.—The given diagonal must be greater than the side, and less than the diagonal of the given square.
Note 2.—The given base must be less than the diagonal of the given square.

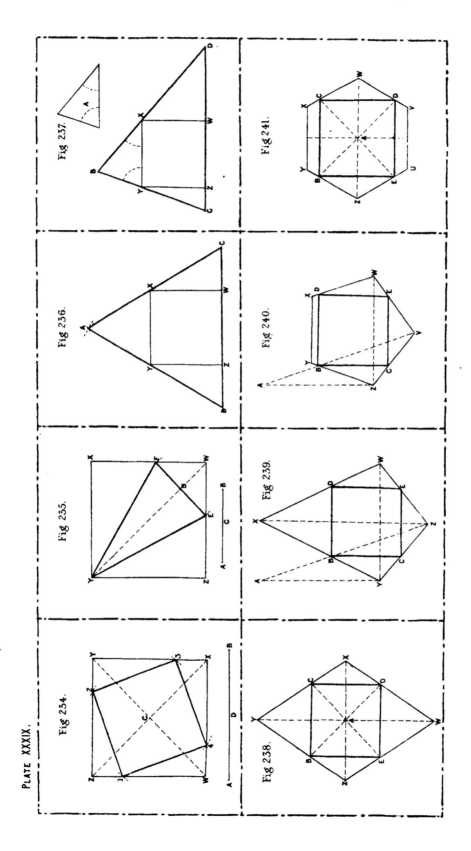

PLATE XXXIX.

Fig 234.

Fig 235.

Fig 236.

Fig 237.

Fig 238.

Fig 239.

Fig 240.

Fig 241.

FIG. 242.—*To inscribe a hexagon in a given square*, Z Y X W.

Draw the diagonals crossing in A,—from A, with any radius, describe a semicircle cutting a diagonal, Y W, in C and B,—from C and B, with the same radius, mark points D and E on the semicircle,—draw lines from A through D and E, producing them both ways to cut the sides of the square in points F, G, H, and I,—from A, with radius A G, cut Y W in K and L,—join points F, G, K, H, I, and L.

FIG. 243.—*To inscribe an oblong in a trapezion*, Z Y X W, *one side*, V U, *of the oblong, being given*.

Draw the diagonals crossing in A,—bisect line V U in B,—from A, with radius V B, mark on the shorter diagonal points C and D,—through C and D draw lines parl. to the longer diagonal, to cut the sides in E and F, and G and H,—join E G and F H. (See Note.)

FIG. 244.—*To inscribe a rhombus in any given rhomboid*, Z Y X W.

Draw the diagonals of the rhomboid crossing in A,—through A draw lines bisecting the angles formed by the diagonals at A, producing them to cut the sides of the rhomboid in points B, C, D, and E,—join these points.

FIG. 245.—*To inscribe a pgram. in any given trapezium*, Z Y X W.

Bisect the sides of the trapezium in points A, B, C, and D, and join these points.

FIG. 246.—*To construct an eqtrl. triangle about a given circle, or to inscribe one within a circle*.

Through the centre of the circle, Z, draw a diameter, A B,

—from A and B, with the radius of the circle, mark points C, D, E, and F,—join the alternate points, D, A and E,—or,—through the points D A and E, draw tangents to the circle, to meet in points G, H, and I.

(fig. 160)

FIG. 247.—*To inscribe within a given circle*, Z, *a triangle similar to a given triangle*, Y.

At any point, A, in the circle, draw a tangent to it,—at A, on either side of A, construct two angles with the tangent, equal two of the angles of the given triangle, —produce the sides of the angles to cut the circle in points C and B,—draw line C B.

FIG. 248.—*About a given circle*, Z, *to construct a triangle similar to a given triangle*, X Y W.

Draw a radius, Z A, to the given circle,—produce a side of the triangle both ways, as O X and W B,—at the centre, Z, on either side of Z A, construct angles equal the angles O X Y and Y W B,—produce the sides of the angles to cut the circle in points D and E,—through D, E, and A, draw tangents to the circle, to meet in points F, G, and H.

FIG. 249.—*To construct a square about a given circle, or to inscribe a square within it*.

Draw two diameters to the circle, perpr. to each other, cutting the circle in points A, B, C, and D,—through A, B, C, and D, draw tangents to the circle, meeting in points E, F, G, and H,—and join the points A, B, C, and D, for the inner square.

Note.—Points O and D must be marked on the diagonal that joins the equal angles, whether it is the longer or the shorter of the two.

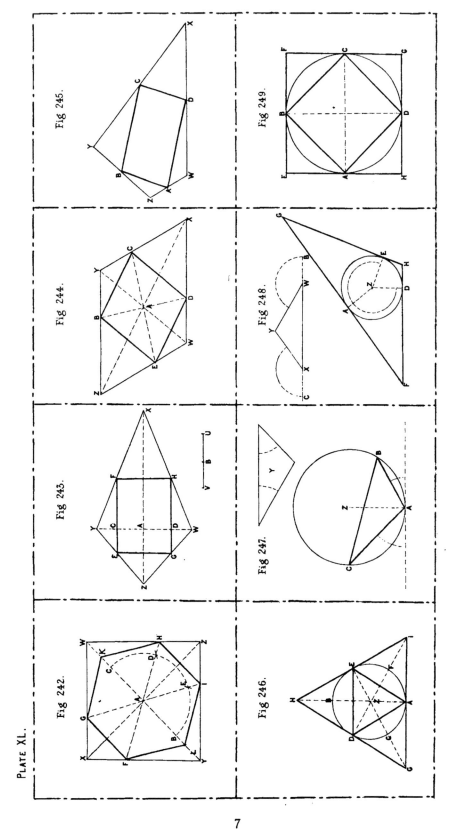

PLATE XL.

Fig 242.

Fig 243.

Fig 244.

Fig 245.

Fig 246.

Fig 247.

Fig 248.

Fig 249.

7

FIG. 250.—*To inscribe a circle within a given rhombus.*

Draw the two diagonals crossing in point A,—from A draw a line, A B, perp. to one side of the rhombus,—from A, with A B as radius, describe the circle.

FIG. 251.—*To inscribe a circle in a given trapezion, Z Y X W.*

Draw the diagonal, Z X, joining the unequal angles,—draw line W A, bisecting the angle Z W X, and cutting line Z X in point A,—from A draw a line, B A, perp. to a side of the trapezion,—from A, with B A as radius, describe the circle.

FIG. 252.—*To construct a rhombus about a given circle, Z, one angle, Y, of the rhombus, being given.*

Draw two diameters, A B and O D, perpr. to each other, producing them outwards,—bisect angle Y,—at Z construct an angle, O Z E, equal half angle Y,—through E draw a tangent to the circle, to cut the diameters produced in. F and G,—make Z H equal Z G,—make Z I equal Z F,—join points F, H, I, and G.

FIG. 253.—*To inscribe three equal circles in an eqtrl. triangle, each circle touching two sides of the triangle, and the two other circles.*

Draw lines bisecting the angles, and sides of the triangle, dividing it into three equal trapezions, as A, B, C, D, &c.,—in each trapezion inscribe a circle. (fig.251)

FIG. 254.—*To inscribe three equal circles in a given eqtrl. triangle, each circle touching only one side of the triangle, and the other circles.*

Draw lines as before, bisecting the angles and the sides, and dividing the triangle into three equal triangles, as A B C, (fig.223) in each triangle inscribe a circle,—if lines are drawn through the centres of these circles, parl. to the sides of these circles, they will cut each other in points 1, 2, and 3, which will be the centres of circles equal to the first three, that will each touch two sides of the triangle, and two of the circles.

FIG. 255.—*To inscribe three circles in an isosceles triangle, Z Y X, each circle touching two sides of the triangle, and the two other circles.*

Bisect Y X in A,—draw line A Z, producing it beyond Z,—draw lines bisecting the angles Z Y A and Z A Y, and crossing in B,—through B draw a line parl. to Y X, crossing Z A in C,—make C D equal O B,—from B and D, with radius B C, describe the two first circles,—at a distance from Z Y equal radius B C, draw a line, E F, outside the triangle, parl. to Z Y,—produce E F to cut Z A produced in G,—through B draw a line perpr. to E F, to cut it in I, and line Z A in H,—from H, with H I as radius, describe an arc,—from G draw a line through B, to cut the arc in K, draw K H,—from B draw a line parl. to K H, to cut Z A in L, and the circle in M,—from L as centre, with L M as radius, describe the third circle.

FIG. 256.—*To inscribe three equal semicircles in a given eqtrl. triangle, Z Y X, having their diameters adjacent, and each semicircle to touch only one side of the triangle.*

Draw lines bisecting the angles of the triangle and the sides in points A, B, and C,—bisect one angle at B, by a line to cut line A X in D,—from E, the centre of the triangle, with E D as radius, mark points F and G,—join points F, G, and D, and these lines are the diameters of the semicircles which will touch the triangle in points A, B, and C.

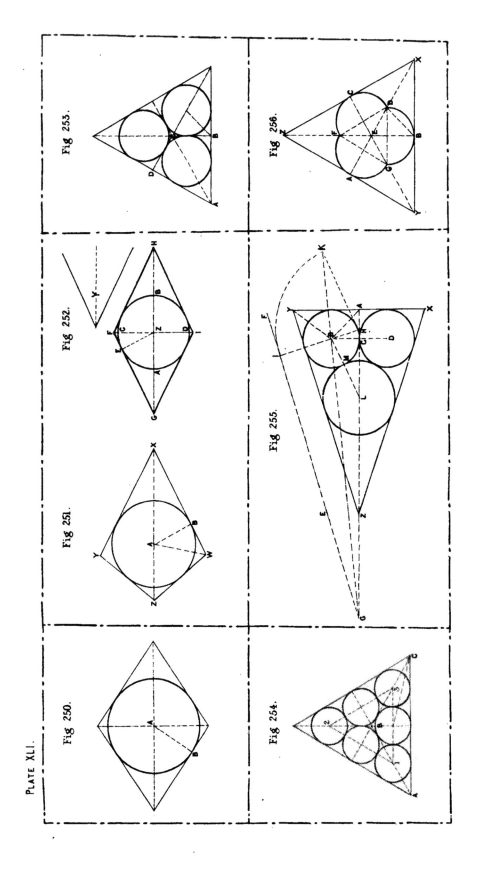

PLATE XLI.

Fig 250.

Fig 251.

Fig 252.

Fig 253.

Fig 254.

Fig 255.

Fig 256.

FIG. 257.—*To inscribe three equal semicircles within a given eqtrl. triangle, Z Y X, their diameters being adjacent, and each semicircle touching two sides of the triangle.*

Draw lines bisecting the angles of the triangle and the sides in A, B, and C,—draw A B, cutting Y C in D,—draw D E, perpr. to Z Y,—from D, with D E as radius, mark A B in points 1 and 2,—from Y through 1 and 2 draw lines to cut lines A X and B Z in F and G,—make C H equal A F,—join points F, G, and H, for the diameters of the required semicircles.

FIG. 258.—*To inscribe four equal circles in a given square, Z Y X W, each circle to touch two sides of the square and two circles.*

Draw the diagonals of the square, and two diameters through their intersection, perpr. to the sides, and bisecting the sides in points A, B, C, and D,—join points A, B, C, and D, by lines cutting the diagonals in points 1, 2, 3, and 4,—these points are centres of the required circles to touch the square.

FIG. 259.—*To inscribe four equal circles in a square, Z Y X W, each circle to touch two others, and one side of the square.*

Draw the diagonals of the square, dividing the square into (fig. 223) four equal triangles,—in each triangle inscribe a circle.

FIG. 260.—*To inscribe four equal semicircles in a square, Z Y X W, each semicircle to touch only one side of the square, and their diameters to be adjacent.*

Draw the diagonals, Z X and W Y, crossing in A,—bisect sides Z W and W X, in B and C,—draw B C, cutting W Y in D,—from A, with A D as radius, mark points E F and G,—the lines joining the points D, E, F, and G, are the diameters of the required semicircles.

FIG. 261.—*To inscribe four equal semicircles in a square, Z Y X W, their diameters being adjacent, and each semicircle touching two sides of the square.*

Draw the diagonals and diameters as before,—bisect an angle formed by a diagonal and one side, by a line produced to cut the opposite diagonal in a point, as A,—from the centre of the square, O, with O A as radius, mark points B, E, and D, on the diagonals,—through these points draw lines parl. to the diagonals, to meet in points O, O, O, O,—these lines are the diameters of the required semicircles.

FIG. 262.—*To inscribe six equal circles in a given circle, Z, each touching the given circle, and two other circles.*

Draw a diameter, A B,—from A and B, with radius A Z, divide the circle into six equal parts, in points C, D, E, and F,—draw radii from C, D, E, and F,—divide a radius into three equal parts, in points 1 and 2,—from Z, with radius Z 1, describe a circle to cut all the radii in points O, O, O, &c., which are the centres, while the distance to the given circle is the radius of the required circles. (See Note.)

FIG. 263.—*To describe six equal circles about a given circle, Z, each circle to touch the given circle and two others.*

As before, draw radii dividing the circle into six equal parts, and produce the radii outwards,—from the centre of the given circle, with its diameter as radius, describe a circle, to cut these radii in points which will be the centres of the required circles, having the same radius as the given circle.

FIG. 264.—*To inscribe any number of equal circles in a given circle, each touching the given circle and two others.*

Divide the circle into the number of equal parts that it is required to contain circles,—draw radii to these points,—bisect one of these arcs, as A B in C,—at C draw a tangent, C D,—from C draw a radius, C Z,—draw a line bisecting angle Z C D, and produce it to cut radius B Z in point E,—from Z as centre, with Z E as radius, describe a circle cutting all the radii in points, which will be the centres of the required circles, with the line E B as radius for each.

Note.—A circle of equal radius may be described from the centre of the given circle, to touch all the six inscribed ones.

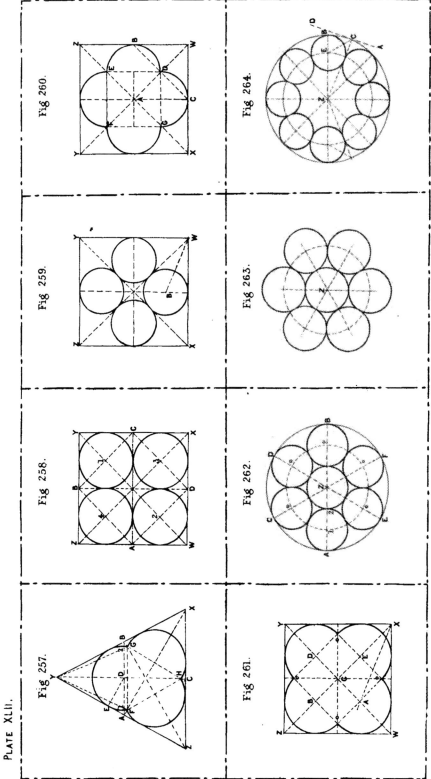

PLATE XLII.

Fig 257.

Fig 258.

Fig 259.

Fig 260.

Fig 261.

Fig 262.

Fig 263.

Fig 264.

FIG. 265.—*To describe any number of circles, about a given circle, each touching the given circle and two others.*

Divide the circle into the same number of parts as are required circles,—draw radii through the points of division, producing them outwards,—bisect one of these arcs, as A B, in O,—draw a radius through O,—at O draw a tangent to cut a produced radius in D,—bisect the outer angle formed at D, as O D E, by a line cutting the radius through O in F,—from the centre of the circle, Z, with Z F as radius, cut all the produced radii in points 1, 2, 3, 4, and 5, which are the centres, and F O the radius of the required circles.

FIG. 266.—*To inscribe any number of equal semicircles, in a given circle, their diameters being adjacent, and the semicircles touching the given circle.*

Draw radii dividing the circle into the same number of equal sectors, that there are required semicircles, and inscribe a semicircle in each sector, by Fig. 224.

FIG. 267.—*To construct any number of eqtrl. triangles above six, about a given circle, touching it, and having their bases adjacent.*

Divide the circle as before, producing the radii outwards,—bisect all the divisions of the circle, as A B in O,—at O draw a tangent,—at O draw a line at 60° to the tangent, to cut a produced radius in a point, D,—from the centre, Z, of the circle, with Z D as radius, describe a circle cutting all the produced radii in points 1, 2, 3, 4, &c.,—join all these points together, and each one to two bisections of the divisions of the circle.

FIG. 268.—*To construct any number of squares above four, about a given circle, their diagonals being adjacent, and each square touching the given circle.*

Proceed as in the last problem, but at the point of contact, O, construct an angle of 45°, instead of an angle of 60°, and

from D, with D O as radius, cut the radius produced through O in E.

FIG. 269.—*To inscribe within any regular polygon a number of semicircles equal to the number of sides, having their diameters adjacent, and each semicircle touching only one side of the polygon.*

Draw lines from all the angles of the polygon to its centre, and inscribe a semicircle in each triangle, as in Fig. 224.

FIG. 270.—*To inscribe in any regular polygon a number of semicircles equal to the number of sides of the polygon, having their diameters adjacent, and each semicircle touching two sides of the polygon.*

Join all the angles to the centre, Z,—bisect all the sides, and join the points of bisection to the centre,—join two of these bisections, as A and B,—upon A B describe a semicircle,—from its centre, O, draw a line perpr. to the side of the polygon, to cut the semicircle in D,—join D to the centre of the polygon, by a line cutting a side in E,—from E draw a line parl. to D O, to cut O Z in F,—through F draw a line perpr. to C Z, to cut A Z and B Z in points G and H,—from F, with F E as radius, describe the semicircle and complete the series.

FIG. 271.—*To describe about or within any given regular polygon, Z Y X, &c., a similar one co-centric with it, having one side, A B or C D, given.*

Draw lines from the centre through the angles of the polygon, as E Z and E Y, &c.,—on the side, or side produced, mark the given side, as Z F, equal A B, or Z G equal C D,—from points F and G draw lines parl. to Z E, to cut line Y E in H and I,—complete the polygons by describing circles with radii E I and E H, to cut the lines drawn through the angles.

FIG. 272.—*To inscribe within any regular polygon a polygon of double the number of sides.*

Draw lines from the centre to the angles,—bisect each of the sides,—from the centre, with the distance to the sides as radius, cut all the radii, and join these points to the points of bisection.

PLATE XLIII.

Fig 265.

Fig 266.

Fig 267.

Fig 268.

Fig 269.

Fig 270.

Fig 271.

Fig 272.

Area, signifies the superficial contents of a figure, or, the amount of surface contained therein.

FIG. 273.—The area of any pgram. is found by multiplying the measurement of the base, by the measurement of the altitude, in similar units of measurement. In the pgram., ABCD, if it is supposed to be 8 inches in base, AB, and 6 inches in altitude, AD, by multiplying the 8 by 6 we have 48 inches, which are called square inches, there being really 48 squares of an inch side. The same would apply to any other units, as feet, or yards: thus, the floor of a room may be said to contain a number of square yards. Of course fractions of the units may be multiplied in the same way.

When we say, a number of feet square, as the area of a figure, it means that the figure is a square of so many feet side, and this distinction must be remembered; for the difference is very great: as for example, 6 feet square, is a square of 6 feet side, and therefore containing 36 square feet, while 6 square feet would only be the contents of a pgram. 3 feet long by 2 feet high.

FIG. 274.—From the above it will be seen that,—

All pgrams. having the same bases and the same altitudes, must have the same areas, and pgrams. having the same base must have to each other the same ratios that their altitudes have; or having the same altitudes, they must have to each other the same ratios, in area, that their bases have.

The pgram, ABCD, is equal to the pgram. ACED.

The triangle, ABC, is half the pgram. ABCD; its area may be found by multiplying the base by half the altitude, or its altitude by half the base, and the same may be said of the triangles AEB and ADB, and the pgram. AEDB, for:—

FIG. 275.—*Triangles are always equal to half the areas of pgrams. of the same bases and altitudes;* from this it will be seen that, *All triangles of equal bases and equal altitudes, have equal areas, and that all triangles of equal bases, have to each other the same ratio*

that their altitudes have, or having equal altitudes, they have the same ratio to each other that their bases have.

FIG. 276.—Exemplifies the last statement; triangle ABC is equal to triangle BDO, their bases and altitudes being equal, and the triangle ACD is divided into two equal parts by the line DE, because AE is half the altitude, AC. (See Note 1, Plate XLV.)

FIG. 277.—The contents of a circle can only be computed in this manner; if we divide it into a great number of equal sectors, by lines from its centre, and suppose these sectors to be triangles, and then add these triangles together, we shall have a pgram. equal to the contents of the circle, practically losing only what would result from the difference between the right line of the base of the triangle and the arc of the sector.

The circle is here shown as equal to a pgram having half its circumference as base, and its radius as altitude, or it is also equal to a pgram. which has its diameter as base, and the quarter of its circumference as altitude.

FIG. 278.—Shows a square, a circle inscribed, and a triangle of the same base and altitude—they are to each other in area in the ratio of 4, 3, and 2.

FIG. 279.—*To find a line equal to the semi-circumference of a given circle.*

Draw a diameter, AB,—draw lines AC and BD perpr. to AB,—make BD equal three times the radius of the circle,—from A, with the radius of the circle, mark point D,—draw a radius bisecting the arc AD, and produce it to cut line AC in E,—join points E and A,—EA is the line required.

FIG. 280.—*To describe a circle, and to construct a square and an eqtrl. triangle each of the same given perimeter, AB.*

For all practical purposes, the diameter and the circumference of a circle, may be said to be in the ratio of 7 to 22 to each other;—divide AB into 22 equal parts, and take 3½ of these parts as radius for the circle,—divide AB into 4 equal parts for the side of the square, and into 3 equal parts for the side of the triangle. (See Note 2, Plate XLV.)

PLATE XLIV.

Fig 273.

Fig 274.

Fig 275.

Fig 276.

Fig 277.

Fig 278.

Fig 279.

Fig 280.

Note 1.—To find the area of any other figure than a pgram., it must be reduced to a triangle of equal area, by geometry, and the contents of the triangle measured; or it may be cut up into a number of triangles, and their contents measured and added together.

Note 2.—The circle is the greatest, and the triangle the least area of the same perimeter. The square is the greatest area of all pgrams, of the same perimeter, and the eqtrl. triangle the greatest triangle of the same perimeter.

FIG. 281.—*To describe a circle equal in perimeter to any number of given circles, AB and BC.*

Place the circles in contact, having a line through all the centres, forming a series of diameters, as AC,—on AC describe the circle required.

FIG. 282.—*To divide the area of a circle into any number of equal or proportional parts, by concentric divisions.*

Draw a radius to the given circle, as AB,—divide AB as you require the area of the circle to be divided, into the number of equal or proportionate parts, as 1, 2, 3,—draw perprs. to AB, from points 1, 2, and 3, to cut a semicircle on AB, in points C, D, and E,—with the centre A describe circles through points C, D, and E.

FIG. 283.—*To divide the area of a circle into a number of parts, equal in area and perimeter.*

Draw a diameter, AB,—divide AB into the number of equal parts by points 1, 2, and 3,—on A1 describe a semicircle,—on 1B describe a semicircle on the opposite side of AB,—on A2 and 2B describe semicircles similarly, and so complete the figure.

FIG. 284.—*To reduce any irregular polygon, as ABCDE, to a triangle of equal area.*

Produce the base, AE, both ways,—join point A to the next point but one, as point C,—from point B draw a line parl. to AO, to cut AE produced in F,—join FO,—similarly, join EC,—from D draw DG, parl. to OE, and join O and G,—FOG is equal in area to ABODE. The same method may be followed over any number of sides, in whatever position, reducing the figure by one side each time.

FIG. 285.—*To construct a triangle equal in area to any regular polygon, as ABODEF.*

Polygon ABODEF is equal to six triangles of AF as base and XY as altitude,—therefore, it is equal to a triangle of six times AF as base, and XY as altitude, or twice AF as base, and thrice XY as altitude.

FIG. 286.—*To construct a pgram. equal in area and perimeter to a given triangle, ABC.*

Bisect AB in D,—produce AC, making CE equal BC,—bisect AE in F,—through C draw a line parl. to AD,—from A and D, with radius AF, cut this line in G and H,—ADHG is the pgram.

FIG. 287.—*To construct an eqtrl. triangle, equal in area to any given triangle, ABC.*

On one side, as AC, construct an eqtrl. triangle, ACD,—produce AD,—draw a line through B parl. to AC, to cut DA produced in E,—find a mean propnl. between EA and AD in AF,—AF is the side of the required triangle.

FIG. 288.—*To construct a square, equal in area to a given pgram., ABCD.*

Find the mean propnl. between the base AB and the altitude BE in BF,—BF is the side of the square equal to the pgram., or in other words, the square upon the mean, is equal to the rectangle contained by the two extremes.

PLATE. XLV.

Fig.281.

Fig.282.

Fig.283.

Fig.284.

Fig.285.

Fig.286.

Fig.287.

Fig.288.

FIG. 289.— *To construct a figure similar to, and equal in area, to any two given similar figures, as* ABC *and* XYZ.

Take any two homologous sides, as AB and ZY, and place them at right angles to each other, as DCE,—line DE, the hypothenuse, is the homologous side to AB and ZY of a figure equal in area to the given figures,—a figure may now be constructed on DE similar to the given figures,—this applies equally to any figure.

FIG. 290.—*To construct a figure similar to, and equal to the difference between, any two given similar figures, as* ABCD *and* ZYXW.

Draw a line, EF, equal to any side of one figure, as DC,—at E erect a perpr.,—from F, with the homologous side of the second figure as radius, describe an arc to cut the perpr. in point G,—a figure constructed similar to the given figure, and with GE as a side homologous to DC, would contain an area equal to the difference in area between ABCD and ZYXW.

FIG. 291.—*To construct a figure similar to a given figure,* ABCD, *and in any proportion to it in area.*

(fig. 24)

Find a mean proportion between any one side, and a line in the same ratio to it, that the required figure is required to be to ABCD, and this mean is the homologous side of a similar figure of the required area,—thus, ZYXW is 2¼ times the area of ABCD, ZW being the mean between AD and 2¼ times AD,—this applies equally to all figures,—for example, the hexagon ZY is ¾ the area of hexagon AB, and the circle Y is 1⅜ times the area of circle X.

FIG. 292.—*To describe a circle to contain a given area, as represented by the square on* AB.

Find a line, ZY, a mean between two lines, in the ratio to each other of 7 to 22, as lines ZX and ZW,—from X make XV equal AB,—and from X make XT equal ZY,—join ZT,—draw a line from V parl. to ZT, to cut ZX in S,—from X, with XS as radius, describe the circle.

FIG. 293.—*Upon a given base,* AB, *or base produced, to construct a triangle equal in area to a given triangle,* ABC.

Join D to the opposite angle of the triangle, as C,—from B draw a line parl. to DO, to cut the side, or the side produced, of the triangle in E,—join points E and D,—the letters apply equally to both figures,—AED is the triangle required.

FIG. 294.—*To construct a triangle equal in area to a given triangle,* ABC, *and of a given altitude,* DE.

At a distance from AB equal DE, draw a line parl. to AB, to cut line BC or BC produced in point F,—draw FA,—from C draw a line parl. to FA, to cut the base AB, or AB produced in G,—join points F and G,—FGB is the triangle required.

FIG. 295.—*On the base, or the base produced, of a given triangle,* ABC, *to construct a second triangle, with its apex in a given point,* X, *and equal to the given triangle.*

Draw a line through A parl. to BC,—draw BX, producing it to cut the line through A in D,—draw XC,—draw a line from D parl. to XC, to cut BC produced in E,—draw line XE,—BXE is the triangle required.

PLATE XLVI.

Fig. 289.

Fig. 290.

Fig. 291.

Fig. 292.

Fig. 293.

Fig. 294.

Fig. 295.

FIG. 296.—*To draw a line to bisect any given triangle, ABC, from any given point, D, on one side.*

Join D to the opposite angle, as C,—bisect A B in E,—draw E F parl. to D C,—join points D and F.

FIG. 297.—*To draw a line to bisect any given triangle, ABC, from any given point, D, within it.*

Draw a line through point D, parl. to the side farthest from D, to cut the side nearest to D in E,—bisect the side farthest from D, and the side nearest in points F and G,—draw E F, and a line from G parl. to E F, to cut A C in H,—from H draw a line parl. to A B, to cut E D produced in I,—describe a semicircle on D I,—on the semicircle from D, with radius D E, mark point K,—from H, with chord K I as radius, mark, past point F, point L,—from L draw the line through point D, to cut the side of the triangle in M,—L M bisects the triangle A B C.

FIG. 298.—*To bisect any given triangle, ABC, by a line parl. to one side.*

Find a mean, B D, between B O and half B C,—from D draw a line parl. to A C, it bisects the triangle A B C.

FIG. 299.—*To bisect any given triangle, ABC, by a line perpr. to one side.*

Produce the side A B,—from C draw a perpr. to A B, to cut it in D,—bisect A B in X,—find A E a mean between A X and A D,—draw E F, perpr. to A B, bisecting the triangle.

FIG. 300.—*To draw lines from the angles of a triangle, ABC, to meet in one point in it, dividing it into three equal parts.*

Bisect two of the sides as B C in D, and C A in E,—draw D A and B E crossing in F,—draw lines from F to the points of the angles.

FIG. 301.—*To find a point in a triangle, ABC, from which lines may be drawn to the angles, dividing the triangle in parts, in a given ratio, as 1, 2, and 3.*

Divide a side of the triangle, as A C, in the required ratio, in points 2 and 3,—from points 2 and 3 draw lines parl. to the opposite sides of the triangle, to meet in D,—from D draw lines to the angles of triangle.

FIG. 302.—*To divide a triangle into any number of equal or propnl. parts, by lines parl. to one side.*

With A B C as the given triangle, divide one side, as A C, as the triangle is required to be divided, in points 1, 2, 3, &c,—on A C describe a semicircle,—from points 1, 2, 3, &c., draw perprs. to A C, to cut the semicircle in points Z Y X W V, &c.,—from point A describe arcs through the points Z Y X, &c.,—to cut A C in points O, O, O, &c.,—from points O, O, O, &c.—draw the lines parl. to the side of the triangle.

FIG. 303.—*To divide a triangle, ABC, into any number of equal or proportionate parts, by lines from a point, D, in one side.*

Divide the side containing point D, as A C, as the triangle is to be divided, in points 1, 2, 3, and 4,—from point D draw a line to the opposite angle, B,—from points 1, 2, 3, and 4, draw lines parl. to D B, to cut the sides of the triangle in points P, P, P, and P,—join points P, P, P, P, to point D.

PLATE XLVII.

Fig.296.

Fig.297.

Fig.298.

Fig.299.

Fig.300.

Fig.301.

Fig.302.

Fig.303.

FIG. 304.—*To divide a triangle, Z Y X, into any number of equal, or propnl. parts, by lines from a point within the triangle, as A.*

(fig. 295)   Find a triangle, Y A B, equal triangle Z Y X,—divide the base, Y B, as it is required to divide the triangle, in points 1, 2, and 3,—from the points within the triangle, as points 1 and 2, draw lines to point A,—from the point A draw a line to point X,—from point 3 draw a line parl. to X A, to cut X Z in C, and similarly from any other divisions outside the triangle, —join point C to A. Should line 3 C go beyond point Z, —line X Z must be continued to meet line 3 C, and a line drawn parl. to a line, Z A.

FIG. 305.—*To add two given triangles together, as Z Y X to A B C.*

(fig. 284)   On a side of the greater triangle, as A B C, construct from one of its angles, a triangle equal the lesser, as B E D,—make the quadrilateral, D F C B, equal the figure B D E A C,— reduce the figure D F C B to a triangle, B G C.

FIG. 306.—*To construct a triangle, equal to a given triangle, Z Y X, and similar to a given triangle, A B C.*

Draw C D, at any angle to B C,—make C D equal the mean between half the base, and the altitude of triangle A B C, —*i. e.*, equal to the side of a square equal to the triangle,—on C D mark C E, equal the mean between half the base, and the altitude of triangle Z Y X,—*i. e.*, equal to the side of a square equal to the area of triangle Z Y X,—join B and D,—draw E F parl. to B D,—from F draw F G, parl. to B A,—triangle F G C is equal triangle Z Y X.

FIG. 307.—*On a given base, A B, and with a given angle, C A B, to construct a triangle equal in area to the given triangle, Z Y X.*

On A B, from one extremity, construct a triangle, A D E, equal triangle Z Y X,—produce A C to cut a line through D, parl. to A B in F,—from E draw a line parl. to B F, to cut F A in G,—draw G B.

FIG. 308.—*To construct a triangle on a given base, A B, with a given vertical angle, C, and of an area equal the square on Z Y.*

(fig. 35)   Upon A B as a chord, describe a segment, A D B, to contain an angle equal the given angle, C,—find Y X, equal a third propnl. to A B, and Z Y,—at a distance from A B equal twice line, Y X, draw a parl. to A B, to cut the arc in E,—draw E A and E B.

FIG. 309.—*To construct a triangle on a given base, A B, the sum of the remaining sides, as Z Y, and the area, as the square of line X V, being given.*

Bisect A B in C, and Z Y in W,—find line V T a third propnl. to A C and X V,—if A C is greater than X V, it will be a third propnl. less,—produce line A B through A,—make C D equal Z W,—make C E a third propnl. to C A and C D, —at a distance from A B, equal line V T, draw a parl. to A B, —from E draw a perpr. to E B, to cut the parl. in F,—from F draw a line, F A, producing it beyond A,—from D and C draw perprs. to E B, to cut F A in G and H,—from H, with radius H G, describe an arc to cut the line through F in K,—draw lines K A and K B.

FIG. 310.—*To construct a triangle of a given area, as the square of line Z Y, one angle, A B C, and the sum of the sides containing it, as X V, being given.*

From B mark B D, equal twice Z Y,—at a distance from B C equal Z Y draw a parl. to B C, to cut B A in E,—triangle E B D has an area equal the square on Z Y,—divide X V in point W, in segments to contain a rectangle equal the rectangle contained under the lines B E and B D,—make B F equal V W, and B G equal X W,—draw G F,—triangle G B F is the triangle required.

FIG. 311.—*On a given base, A B, and with a given angle, C A B, to construct a pgram. of any given area, as the square on Z Y.*

At B erect a line, B D, perpr. to A B, equal line Z Y,—join D A,—draw a line from D perpr. to A D, to cut A B produced in E,—(Note 1), at a distance from A B, draw a line parl. to A B, cutting A C in F,—draw B G parl. to A F.

Note 1.—This is finding a third propnl. to Z Y and A B.

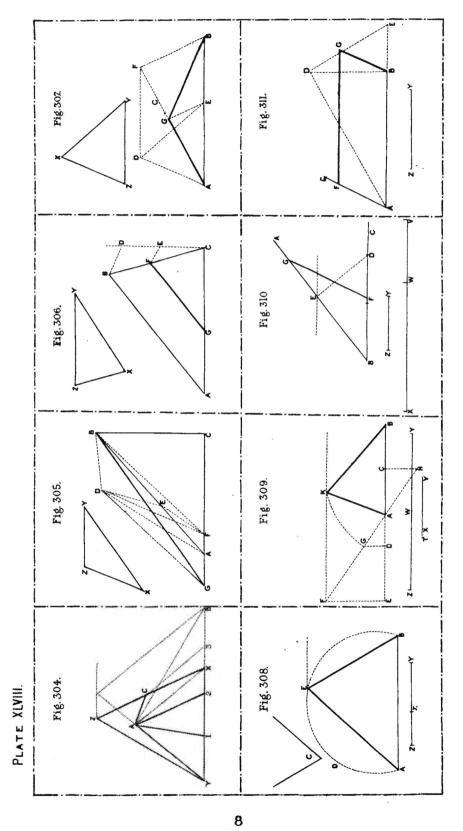

PLATE XLVIII.

Fig. 304.

Fig. 305.

Fig. 306.

Fig. 307.

Fig. 308.

Fig. 309.

Fig. 310

Fig. 311.

PART II.—SIXTH SECTION.—PROBLEMS RELATING TO AREAS.—PLATE XLIX.

Fig. 312.—*To construct a rectangle with sides in a given ratio, as 2 : 3, and of a given area, as the square of Z Y.*

Draw any line divided in the given ratio, as A B in C,—find a mean between the ratios, as C D,—produce C D, and make it equal to Z Y in E,—draw lines from E parl. to D A and D B, to cut A B produced in F and G,—F O and C G are the sides of the required rectangle.

Fig. 313.—*To construct a pgram., similar to a given pgram, A B C D, and of a given area, as the square on Z Y.*

(fig. 288) Draw a diagonal, A C,—from A draw a line at any angle to A D, making it in A E equal to the side of a square equal in area to the pgram., A B C D,—from A mark A F equal line Z Y,—draw line E D and F G parl. to E D,—A G is a side of the required pgram.

Fig. 314.—*On a given base, X Y, or altitude, to construct a pgram. equal to a given pgram, A B C D.*

Produce two sides of the pgram. A B C D as B A and B C,—make a produced part, as A E, equal the given line Z Y,—from E draw a line through D, to cut B C produced in F,—F C is the required side of the pgram.

(fig. 29) Or this problem may be readily accomplished by finding the fourth propnl. to the given base, and the base and altitude of the given pgram.

Fig. 315.—*To construct a rectangle, the area as the square on A B, and the sum, Z Y, of the dissimilar sides being given.*

See problem 30, plate 8.

Fig. 316.—*To construct a rectangle, the area Z Y, and the difference, A B, between the two dissimilar sides being given.*

Bisect A B in C,—at A erect a perpr., A D, equal to Z Y,—from O, with C D as radius, describe an arc to cut A B produced in E,—B E will be one side and A E the second side of the required rectangle.

Fig. 317.—*To bisect a pgram. by a line from any point, Z, in the same plane.*

Draw the diagonals of the given pgram., A B and C D crossing in E the centre,—draw a line from Z through E, producing it to cut both sides of the pgram.

Fig. 318.—*To trisect a pgram. A B C D, by lines from a point, Z, anywhere about the middle of one side.*

Trisect the side containing point Z, as A B, by points 1 and 2,—draw lines through 1 and 2 parl. to the side B C, to cut the side C D in points 3 and 4,—from 3 mark 3 Y equal 2 Z, and from 4 mark 4 X equal Z 1,—join X and Y to Z.

Fig. 319.—*To divide a given pgram, A B C D, into any number of equal parts by lines from one angle, as A.*

If the pgram. is to be divided into an even number of parts, divide each of the sides, opposite angle A, into half the required number of parts; but if an uneven number, as five, divide each side into five equal parts, and draw lines from each alternate division, commencing with the first on either side from point O.

PLATE XLIX.

Fig. 312.

Fig. 313.

Fig. 314.

Fig. 315.

Fig. 316.

Fig. 317.

Fig. 318.

Fig. 319

PART II.—SIXTH SECTION.—PROBLEMS RELATING TO AREAS.—PLATE L.

FIG. 320.—*To divide a pgram., ABCD, into any number of equal parts, by lines parl. to a diagonal, DB.*

(fig.302) Divide the triangle, DBC, into half the number of parts that are required, by lines parl. to the line DB, or, if an un-even number is required, divide the triangle into the full number of parts, and use each alternate division, commencing with the first from DB,—draw a line perpr. to DB, and mark similar divisions on triangle ADB.

FIG. 321.—*To divide a pgram, ABCD, into any number of equal or propnl. parts, by lines from a point, Z, in one side.*

(fig.284) On the base, DC, produced, construct a triangle, ZYX, equal in area to ABCD,—divide YX into the parts that it is required to divide the pgram., in points 1, 2, 3, and 4,—from the points on the side of the pgram. draw lines to point Z, and for any points on the production of the side, as point 1, draw DZ, and a line from 1 parl. to DZ, to cut the side of the pgram. in point E,—join E to Z.

FIG. 322.—*To add any two or more dissimilar pgrams. together, as ABCD and ZYXW.*

(fig.29) Find ZE, a fourth propnl. to the base, ZY, and the base and altitude of the pgram., ABCD,—at a distance from ZY equal ZE draw a line parl. to ZY, to cut the sides XZ and WY produced in F and G,—FGWX is the pgram. equal to the two, and any number of pgrams. may be added in a similar manner.

FIG. 323.—*In a given circle, C, to inscribe a rectangle to contain a given area, as the square on AB.*

(fig.28) Draw a diameter, DE,—find FG a third propnl. to DE and AB,—at a distance from DE equal to FG, on either side of DE, draw two lines parl. to it, to cut the circle in opposite points, H and I,—join points H and I to points D and E.

FIG. 324.—*To find the area of any given trapezium, ABCD.*

Draw a diagonal, AC,—through one angle, as D, draw a line parl. to AC,—from the opposite angle, as B, draw a line perpr. to AC, to cut the line through D in E,—bisect EB in F,—find GC a mean between FB and AC,—GC is the side of a square equal to ABCD in area.

FIG. 325.—*To bisect a trapezium, ABCD, by a line from one angle, as D.*

Draw the diagonals AC and BD,—bisect AC in E,—from E draw a parl. to DB, to cut the side opposite D in F,—join F to D.

FIG. 326.—*To change any rectilinear figure, ABC, to one of greater number of sides, and an equal area.*

Take a point, as D, outside the figure,—join D to one angle near it, as C,—from B, the next near angle to D, draw a line parl. to DC, to cut the base in E,—figure AEDB will be equal to the triangle ABC. To increase by another side, take a point, F, outside BD,—join F to B, and proceed as before, getting figure AGFDE equal to the triangle ABC.

FIG. 327.—*To construct any regular polygon of a given area, as the square on ZY.*

Construct a regular polygon of the required number of sides, as ABCD, with a centre, F,—find line EF, equal to the (fig.29) side of a square, equal in area to the hexagon ABCD,—draw FG, the apothem of the polygon,—find FH a fourth propnl. to EF, FG, and ZY,—FH is the apothem of the required polygon.

PLATE L.

Fig. 320.

Fig. 321.

Fig. 322.

Fig. 323.

Fig. 324.

Fig. 325.

Fig. 326.

Fig. 327.

FIG. 328.—*To construct any multilateral figure, similar to a given figure, and of a given area.*

(fig.29) Let A B C D E F be the given figure, and Z Y the side of the square representing the area,—find line X W, the side of a square equal to A B C D E F in area,—find A G a fourth propnl. to X W, A F, and Z Y,—on A G construct a figure (fig.115) similar to A B C D E F.

FIG. 329.—*To divide any irregular or regular figure, as A B C D E F, into any number of equal or propnl. parts, by lines from any point, P, within it, or on one of its sides, or at an angle.*

(fig.284)
(fig.295) Produce one side of the figure, as E F, and upon it construct a triangle, P Q R, with its vertex in point P equal in area to the given figure,—divide Q R into the number and ratio of parts 1, 2, 3, and 4, as it is required to divide the polygon,—draw line 2 P,—as point 3 falls without the polygon draw line 3 X parl. to line E P,—draw point X to point P,—from point 4 draw a line parl. to 3 X, to cut E D produced in G,—draw G Y parl. to D P,—join Y to P,—draw a line from R parl. to 4 G, to cut E D produced in H,—draw a line from H parl. to G Y, to cut D C produced in I,—draw I Z parl. to C P,—draw Z P, and proceed in the same way with point I. This is in reality an amplification of figure 321.

FIG. 330.—*To construct any eqtrl. triangle equal in area to a given triangle, Z, a pgram., Y, and a pentagon, X.*

(fig.284)
(fig.305) Reduce the pgram. Y and the pentagon X to two triangles,
(fig.287) A B C and D E F,—add the three triangles together in triangle G H I,—construct an eqtrl. triangle, O P Q, equal G H I.

FIG. 331.—*To construct a rectangle, on a given base, A B, equal to three eqtrl. triangles, Z, Y, and X.*

Find three squares, W, V, and U, each equal to a given triangle, by finding the mean between half the altitude and the (fig.289) base,—construct a square, C D E F, equal to the three squares W V U,—find a third propnl. G H, to A B and G D,—on A B construct a pgram. with G H as its altitude.

FIG. 332.—*To construct a pentagon, equal in area, to the difference in area, between a given triangle, Z, and a hexagon, Y.*

Find the mean propnl. A B, to the base and half the altitude, which is the side of a square equal Z,—find a mean propnl. to the apothem of the polygon, and a line equal half its perimeter, as D C, which is the side of a square equal in area (fig.290) to the polygon,—find the side, E F, of a square equal to the difference between the square on A B and the square on D,— (fig.327) construct a pentagon equal in area to the square on E F.

FIG. 333.—*To construct a triangle, a square, and a rhombus, together equal in area to a given polygon, A B C D.*

Construct any triangle, Z, a square, Y, and a rhombus, X,— find W V a side of a square equal triangle Z,—and U T a side of a square equal rhombus X,—and R S a side of a square equal to the three squares, W V, V, and U T,—find a line Q P, the side of a square equal in area to the given polygon, A B C D, by reducing the polygon to a triangle, and finding the side of a square, equal in area to the triangle,—find O N a fourth propnl. to R S, and Q P, and the side of the triangle O N is the side of the required triangle,—find M L, a fourth propnl. to R S, Q P, and the side of square Y,—M L is the side of the required square,—find K I, a fourth propnl. to R S and Q P, and the line U T,—and K I is the side of the required rhombus, similar to rhombus X.

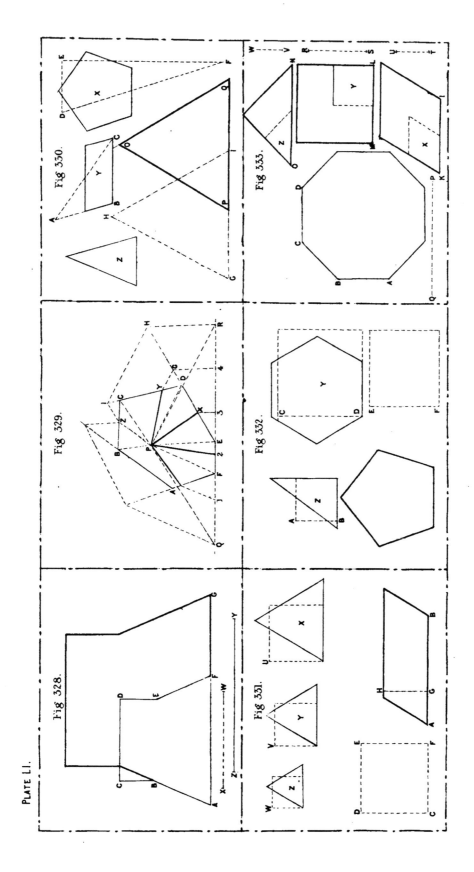

PLATE LI.

Fig 328.

Fig 329.

Fig 330.

Fig 331.

Fig 332.

Fig 333.

PART III.—APPLICATIONS OF GEOMETRY.—Plate LII.

Plate 52 shows a number of examples of surfaces covered by the repetition of geometric figures, and parts of the same figures. As the construction of each of these figures has already been given, the plate will in many respects speak for itself as to construction; it may be suggestive to the student, as an application of Geometry, and, while it acts as an incentive to him to exert his ingenuity in applying the knowledge he has already obtained, the accurate reproduction of these, or similar figures, will be a capital exercise in the use of geometrical instruments.

FIGS. 1, 2, 3, 4, 5, 6, and 8 are sufficiently explained by the working lines that are left upon parts of the figures.

FIG. 7, which resolves itself into the problem, to cover a given space with a series of eqtrl. triangles, squares, and hexagons; the space must first be divided into a series of eqtrl. triangles, as ABC, then bisect the triangle by an altitude, as AD,—bisect the angles ADC and ACD by lines cutting in E, and the distance from A to E is the radius for all the circles to be described cutting each other, in points, which, when joined by right lines, give us the figures required.

FIGS. 9, 10, 11 are self-evident in their construction.

FIG. 12.—Here we have first to divide the space into a series of oblongs, as ABCD,—draw a diagonal, AC, and bisect it in E,—the lines perpr. to and bisecting the lines AE and EC will give us the centres for the curves, when they cut the lines AB and CD, produced if necessary.

PLATE LII.

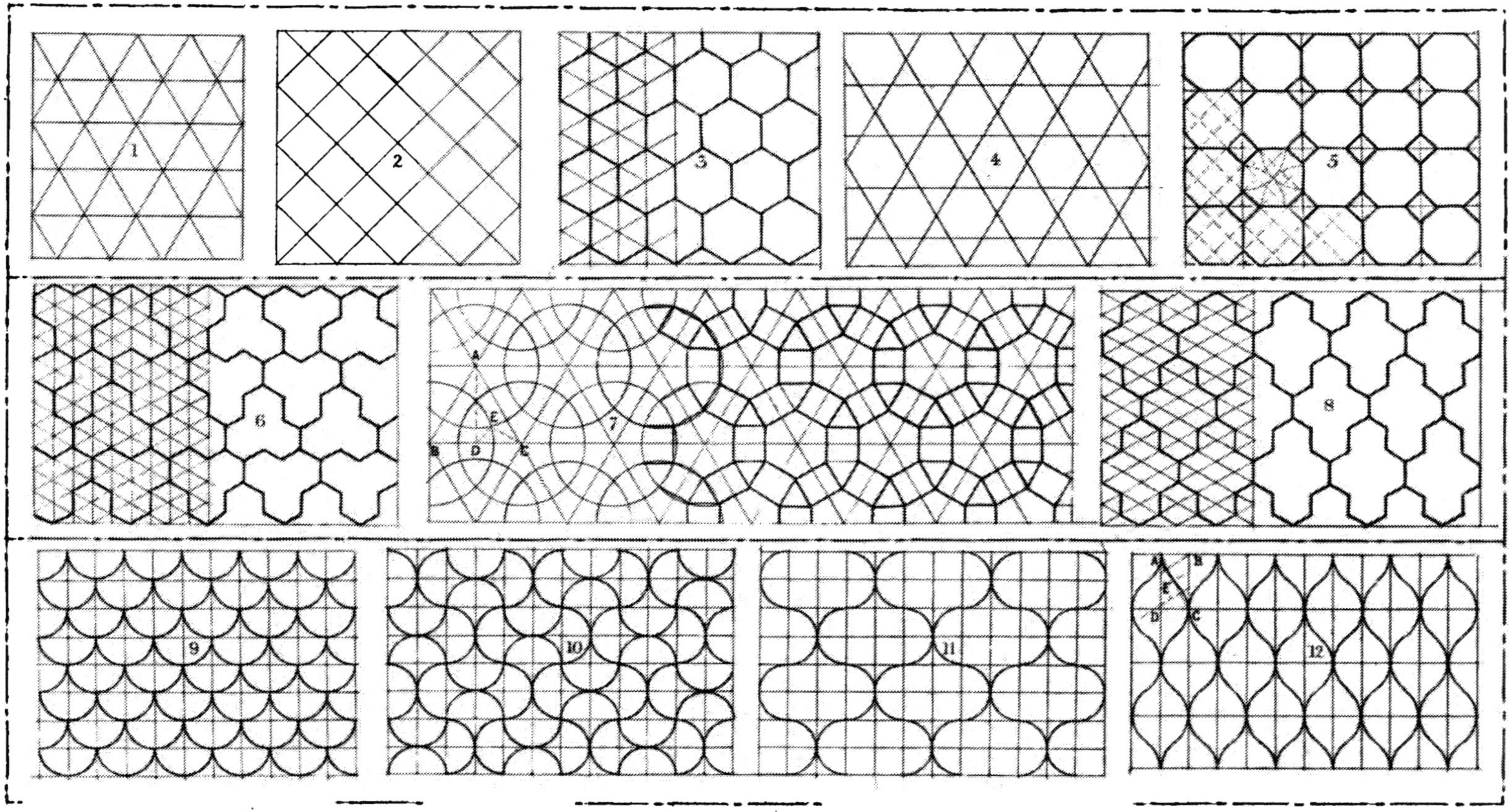

PART III.—APPLICATIONS OF GEOMETRY.—Plate LIII.

In the accompanying plate of arches, it must not be imagined that any set rules are given by which the centres of the curves may invariably be found, but rather a series of suggestions which may be applied according to circumstances, and taste, since there must be considerable variation in many cases, from the different data of height and width.

Fig. 1.—Is called a segmental arch, points A and B called the springing, and point C called the crown, being given; it is simply a segment of a circle described through those points.

Fig. 2.—Is a semicircular arch; stilted, it is termed, because the arc is continued on each side by a short right line.

Fig. 3.—Is a lancet arch, points A, B, and C, being given,—join A and C, and bisect AC by a perpr., producing it to cut a line, A B, produced, in D, which is the centre of the curve.

Fig. 4.—Is called an equilateral pointed arch, from the fact that the distance, AC, is equal to line AB,—points A and B are the centres.

Fig. 5.—Is termed a horse-shoe arch, common in all Saracenic architecture; in this the centres of the curves are always above the line of the springing. It is sometimes round at the top, though in this case it is somewhat pointed; points A, B, and C being given, and AC being bisected by a perpr., the centre, D, is taken on that line, according to the amount of curvature that may be desired.

Fig. 6.—Is an elliptical arch, or, in other words, a semi-ellipse.

Fig. 7.—Is a four centred arch, the points A, B, and C, being given, the arc AC has two centres, i. e., it must be a continuous curve described from two different centres, for this it must be remembered, as in previous curves, the two centres must lie on one line drawn from the point of junction of the curves; the first centre, D, may be any suitable point selected on line A B, and a curve described as from B, as far as desir-able, to point E,—E and C are joined, and E C bisected by a perpr., which is produced to cut line E D produced, in F, which is the second centre from which arc E C may be described in continuation of arc B E.

Fig. 8.—The varieties of the four centred arches are unlimited; in Tudor Architecture we find the first curves from the springing, continued by one, or two right lines meeting in the point C.

Figs. 9 and 10.—These are called ogee arches; the points A, B, and C, being given, the centres DD, and the points of junction, must be determined according to taste, and the necessities of the data, remembering as before, that the centres must fall on one line through the junction.

Figs. 11, 12, and 13.—Are varieties of rampant arches, the distinction being, that the two points of springing are on different levels.

Fig. 11.—A pointed rampant arch is found in a similar manner to the four centred, or pointed arches, the points A, B, and C, being given, and the centres varied according to requirements.

Fig. 12.—An elliptic rampant arch may be constructed by Fig. 117, Plate XX.

Fig. 13.—Is a similar arch to 12, but is approximately found by continuous arcs of circles.

Fig. 14.—Is an example of the trefoil arch, of which there are many varieties, some being pointed, and others round; in this case, the points A, B, and C, being given, line A C is drawn, and the point D is selected; line D C is bisected by a perpr., upon which the centre for the arc D C is taken, and arc D A is described from a point found on line A B, by bisecting A D by a perpr.

Note.—These examples, as well as those of succeeding plates, are given only as indications for the student, of the application of geometry to the various subjects, and not with any idea of their complete treatment.

PLATE LIII.

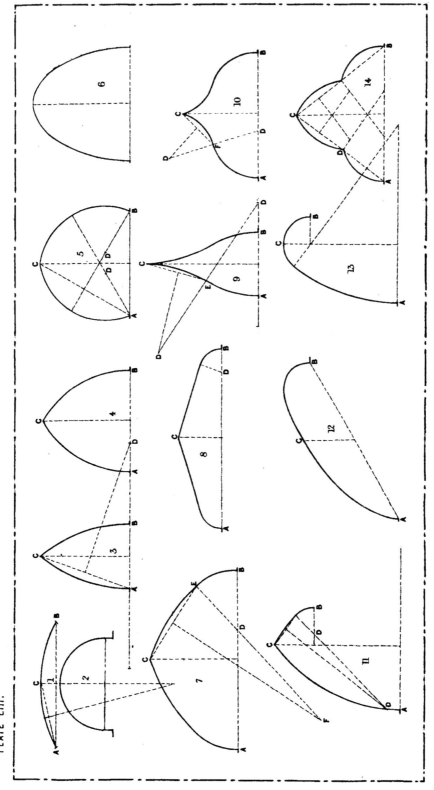

On this plate are shown a series of profiles of classic mouldings, constructed by geometric lines. As a broad distinction, it may be observed, the Roman are composed of curves of circles, while the Greek are all elliptic curves.

FIG. 1—The ovolo, is a simple quadrant.

FIG. 2—Has points A and B given, and the centre for the curve is found by joining A and B, and bisecting A B by a perpr., produced to cut the horizontal through A in C.

FIGS. 3 and 4—The cavettos, are simply 1 and 2 reversed.

FIG. 5—The torus, is a semicircle.

FIG. 6—A cyma recta, or ogee, is composed of two quadrants, having their centres on the same line, A B.

FIG. 7—Has the points A and B given; the centres are found by bisecting the line A B in C, by a horizontal, D E, and bisecting the lines, A C and B C by perprs., produced to cut D E in F and G.

FIGS. 8 and 9—Are the reverse of Figs. 6 and 7, the perprs. to A B in Fig. 9 being produced to cut the horizontals through A and B; this moulding is called the cyma reversa.

FIGS. 10, 11, and 12—Show methods for finding the Greek scotia; points A and B being given, in Fig. 10, it is evident that the ordinates of the semicircle, as line 1, 2, is measured on the horizontals from the same point, as 1, 3. In Fig. 11, the centre for the arc is varied at pleasure, so changing the character of the curve; fig. 12 is the same method as Fig. 117, Plate XX.

FIGS. 13 and 14—Are approximate methods for obtaining the curves of the Greek cavetto, by arcs of circles.

In Fig. 13 the points A and B being given, A D is made two-thirds of the projection A C,—A E, a vertical from A, is made equal to A D,—point E is the centre for one curve, and a line drawn from D, through E, and produced to cut a horizontal through B, in F, gives us the centre for its continuation; this profile may of course be treated as a quarter ellipse.

In Fig. 14, the point C is selected according to taste, and joined to points A and B.

FIGS. 15, 16, 17, 18, and 19—Show approximate methods for obtaining the curves of the scotia, and ovolo, of various proportions, the working lines on the plate being sufficient for the student who has studied the previous parts of the book. In 15 and 16, line C D must be parl. to a tangent to the ellipse, i.e., line B C must be equal to D A : D having been chosen according to the curve required. In 17 and 18, A C is made one-fourth of the height from A to B, and the same distance set up from A to B.

FIGS. 20, 21, and 22—Are examples of the cyma recta and cyma reversa. All these are evidently but compounds of the methods used in Figs. 10, 11, and 12. It must be understood that in these figures the points A and B in the relative position of the one to the other are assumed.

PLATE LIV.

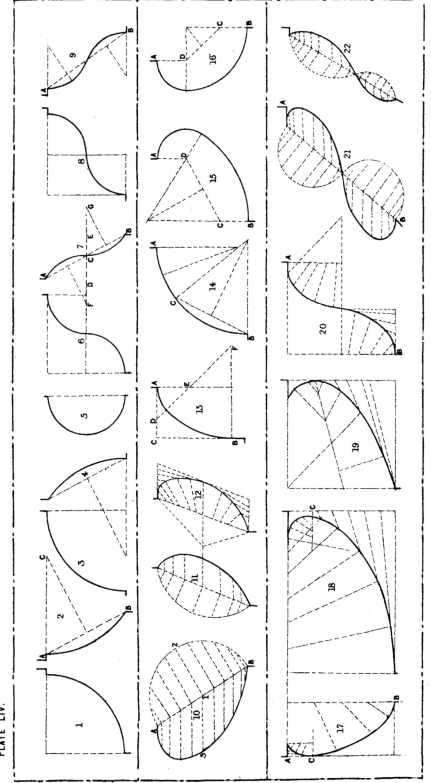

PART III.—APPLICATIONS OF GEOMETRY.—TREFOILS.—PLATE LV.

From the unlimited field of geometric design in Gothic tracery, the accompanying examples are taken, as showing the foundations of many of the most usual figures, with the direct application of the study of geometry.

Fig. 1.—The simplest kind of trefoil may be described as about an eqtrl. triangle, or within one; in the latter case, it is only three equal circles inscribed in the triangle, and stopped at their points of contact.

Fig. 2.—In this case the trefoil is pointed, the three circles inscribed within the triangle being continued, as it were, into the very points of the triangle. Line A B is drawn from an angle through the centre of the circle, cutting it in point C,—line C D is then drawn, and bisected by a perpr., which in its intersection with A B gives the centre for the continuation of the curve. (Note I.)

Fig. 3.—Here we have what may be termed a double trefoil; it is exactly equal to the six equal circles in a triangle, of Fig. 254, Plate XLI, with the unnecessary parts of each circle obliterated.

Fig. 4—Is a concave eqtrl. triangle, turned into a sharply pointed trefoil, by what are termed cuspings; the curves of the triangle may be described from centres found by joining the angles of the triangle, as A B, and bisecting A B by a perpr.; the centre C is chosen, remembering that the distance C D must be equal D E.

Fig. 5.—Shows a more delicately curved trefoil than either of the preceding, the curve being composed of three continuous arcs instead of two. The triangle being given, the centre, A, for the first part of the curve, is first chosen, and the arc B C described; afterwards, the curve is continued from the centres, D and E, which may be varied according to requirements, as in the ogee arches, 9 and 10, Plate LIII.

Fig. 6.—An eqtrl. triangle may be composed of arcs described from its angles, in which case it is called a spherical triangle; and this may be formed into a trefoil by cuspings, similarly to Fig. 4; the distance A B is a matter of choice.

Fig. 7.—Three equal circles may be approximately inscribed in a spherical triangle, to form an included trefoil, by the following means. Draw the three lines bisecting the angles and sides of the triangle, producing one a little beyond it, as at A,—trisect the arc, A B, in points 1 and 2,—at 2 draw a tangent to the curve, continuing it to cut the line produced through A, in C,—bisect the angle A C B by a line, which cutting the line that bisects the angle of the triangle, gives us the centre of one of the circles.

Fig. 8—Which may be termed a trefoil tournante, is a very decided variation upon any of the others; it is inscribed in a spherical triangle, by lines as follows: Draw lines bisecting the angles of the triangle, and lines joining the angles of the triangle, giving the points A, B, and O,—from these points, with the line B D as radius, describe the quadrants, as D E,—from the centre of the triangle, describe the circle, through point E,—divide the arc, F G, into four equal parts,—from 1, and D, with B D as radius, describe arcs cutting in H,—H is the centre for arc D I.

Note I.—As in the plate of arches, the constructions in these figures are only given to one part, or section, the repetition being considered as quite within the scope of any student who has studied the Geometry.

PLATE LV.

Fig 1.

Fig 2.

Fig 3.

Fig 4.

Fig 5.

Fig 6.

Fig 7.

Fig 8.

FIG. 9.—Is composed of four tangential circles; it may be inscribed within a square, or about it, as in this case. These are sometimes spoken of as foils of certain diameters; it will be observed that the side of the square A B is equal to the diameter of each foil.

FIGS. 10 and 11—Require but little explanation; in one, the bisection of the half diameter, and in the other, the bisection of the half side, gives the centre for the opposite curve, though these centres may be varied at pleasure.

FIG. 12.—In this pointed quatrefoil we find point D by the bisection of angle A B C, and continuing the side of the square; the arc E F may be described, when the bisection of E C by a perpr. produces the centre G for the arc E C.

FIG. 13.—This is called a quatrefoil of semicircles.

FIG. 14.—In this figure the diameter of the eye, or centre circle, may be varied to a certain extent, after which A B is bisected for the centre of arc C D,—D E being bisected by a perpr., gives us in F the centre for arc D E,—arc E G is described with the same radius as arc D C.

FIG. 15.—Which is a square pointed quatrefoil, may be considerably varied; here we have the lines of two variations shown, as they may very commonly be found in ornamental stone work. Lines A B and C D are divided into eight equal parts, and by the cross lines we obtain the centres a, a, of arcs b, and i, &c.; centre e may be used for the small arc f.

FIG. 16.—Here we have a section of a group of pillars and mouldings, in which the same principle as the quatrefoil is used, it being based in construction, upon the lines of four equal circles, equidistant from a common centre; the proportions of the various parts are matters of taste, and bearing this in mind, the working lines will be sufficiently explicit.

PLATE LVI.

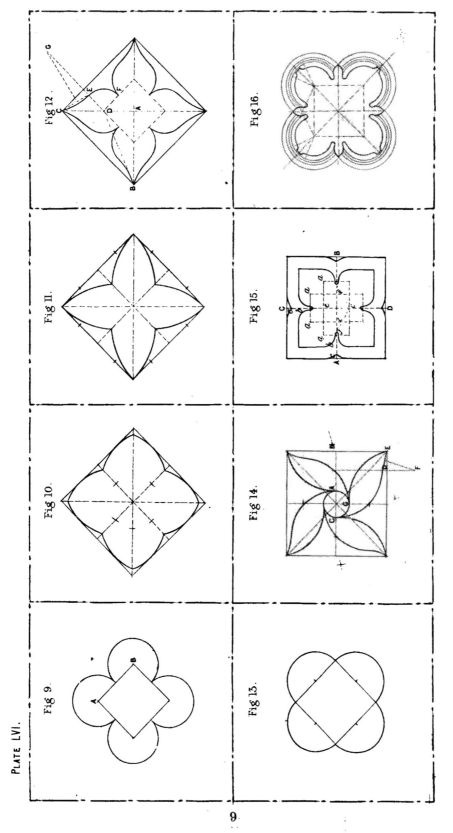

Fig 9.

Fig 10.

Fig 11.

Fig 12.

Fig 13.

Fig 14.

Fig 15.

Fig 16.

The circle may be filled with an immense variety of ornamental geometric forms, of which this plate gives a few examples.

FIGS. 17 and 18—Are really, in each case, only a number of equal circles touching each other, and the including circle, with a portion of each circle left out.

FIG. 19—Which is called a rose tournante, indeed, most of the figures on this plate are called roses, is very simply drawn, all the radii of the curves being the same; the centres 1, 2, 3, and 4 are found by making B 1 equal B D, &c., line 1 A being the radius for the curves.

FIG. 20—Speaks for itself.

FIG. 21.—It should be observed in this figure, that at the points A, A, the curves are only approximately continuous, for it is impossible for the two inner circles to touch each other at both these points without cutting. To construct this figure, the circle B B is described, of a convenient radius,—from the point C, with the radius of the largest circle, an arc is described, and from the centre of the circle B B, with a radius equal to the radii of the two circles added together, a second arc is described, to cut the first in a point, as D. Similarly, the point E is found, the smallest circle being described at choice, though the space Z Y should be nearly, or quite equal to C X.

FIG. 22.—This kind of figure may be constructed with six points, but the proportions are more uniform and graceful when only five are used. The first step in the construction, is to divide the circle into ten equal parts, as 1, 2, 3, 4, &c, and from these points to draw lines to the centre O,—line 1 O is bisected in A for the centre for arc 1 B,—line B 3 is bisected by a perpr., to cut A B produced, in D, as a centre for arc B 3,—and line 3 O is bisected by a perpr., to cut line 2 O in E, as a centre for arc 3 O.

FIG. 23.—Here we have what may be termed a double trefoil; the construction is very simple and intelligible. Point A is the centre of the larger circle, and the bisections of lines A B and A C are the centres of the smaller circles.

FIG. 24.—It will scarcely be necessary to give any further description of this construction than that indicated by the dotted lines, but the student may observe at the beginning, that the same radius must be used for all the curves, great or small, i. e, line A B.

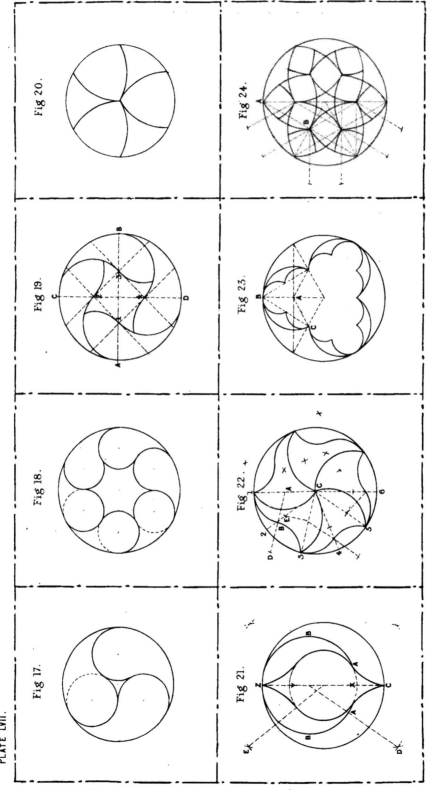

PLATE LVII.

Fig 17.

Fig 18.

Fig 19.

Fig 20.

Fig 21.

Fig 22.

Fig 23.

Fig 24.

Figures may be constructed upon any of the regular polygons as foundations, in exactly the same way as in the trefoils and quatrefoils, but the variety is so large, that in this case, only one is constructed upon each of the most usual polygons, their further amplification being left to the ingenuity of the student.

Fig. 25—The cinq-foil, is very much used; it may be constructed upon a pentagon, as in this case, or within it; in this case the side A B is made equal half the side C D.

Figs. 26 and 27—Show other varieties of the polyfoil.

Fig. 28.—In this example, which shows a polyfoil constructed within a concave octagon, we have the centre of the arc of concavity, A, and the centre of each foil, B, chosen as a matter of taste and convenience, within certain limits,—after drawing the lines A B throughout, the construction is evident.

Fig. 29.—By the inscription of three equal circles within a spherical triangle, each circle touching the other two, we have the foundation of this figure; the centre, C, for the connection and continuation of these circles may be accurately found by finding a curve, equidistant from the circle and the curve of the triangle, to cut the line bisecting the angle.

Fig. 30.—A very much talked of figure, called the pentacle, formed by drawing all the diagonals of a pentagon without the sides, is here a foundation for an ornamental figure; the circles may be made to touch the outer circle, if required. A similar figure may be constructed by joining all the shorter diagonals of a hexagon, and describing circles between the points.

Fig. 31.—This somewhat curious figure is founded on the diagonals of a hexagon, the first step being to divide the diagonal into six equal parts, as 1, 2, 3, 4, 5, by which we obtain the start points for the internal lines, as drawn; the next step is to divide the line A B into three equal parts, in 1 and 2,—points 1 and 2 are centres, as also points A and B, for which curves, it is evident; point C is the centre for arcs X and Y.

Fig. 32.—The construction of this figure, which is like many figures of a much more complicated character, founded upon the simple form of the square, will be evident from the lines on the plate, when it is observed that the points A and B are chosen; after which line A B is divided into ten equal parts.

Note.—The last three figures are taken from examples of the sections of columns and mouldings.

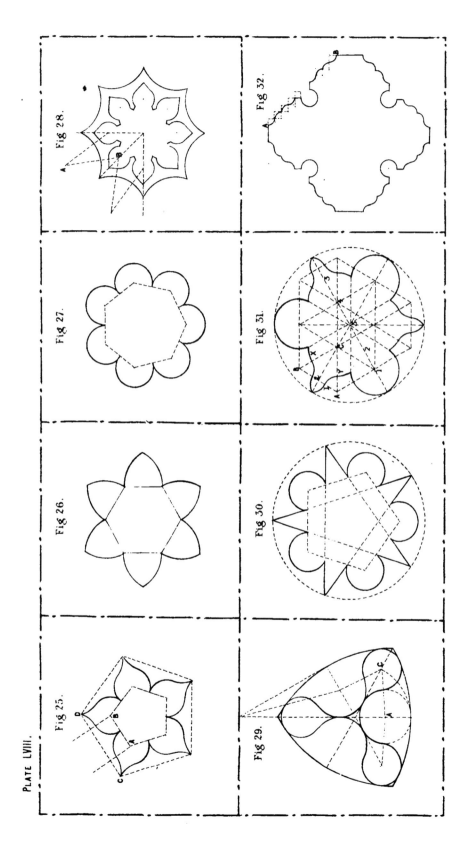

PLATE LVIII.

Fig 25.

Fig 26.

Fig 27.

Fig 28.

Fig 29.

Fig 30.

Fig 31.

Fig 32.

PART III.—APPLICATIONS OF GEOMETRY.—Plate LIX.

For the student who has gone through the previous figures in this part, it will be quite unnecessary to give a detailed description of the examples on this plate, since he will readily perceive, that in the filling in of these arches, he has only an application of what has been before described. In these combinations it must be understood, that only the foundation lines are here indicated, as each space, circle, triangle, or otherwise, could be filled by more detailed forms, and indeed these again might be filled, by even more detailed ornament.

In these figures the lines have been indicated to show some of the centres, and the methods of finding them, but in many instances of finding a point for a centre, equidistant from two or more curves, it is really more practicable to do it by a number of trials or guesses, than by strict geometric construction; but at the same time, the student will be much assisted in his guesses by applying his geometrical knowledge, either mentally or otherwise.

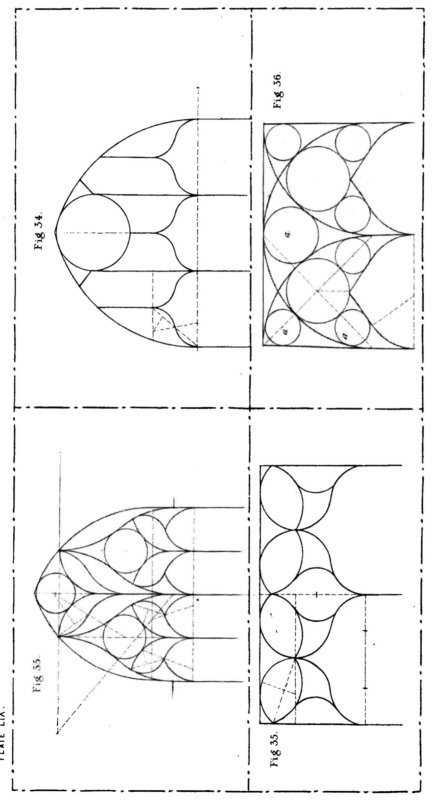

PLATE LIX.

Fig 33.

Fig 34.

Fig 35.

Fig 36.

PART III.—APPLICATIONS OF GEOMETRY.—PLATE LX.

Similarly to Plate LIX, figures 37 and 38 are examples of the application of geometric lines to window tracery. Both these examples are from actual windows, figure 37 being from a Continental church, and figure 38 being from the Archbishop's Palace at Lambeth.

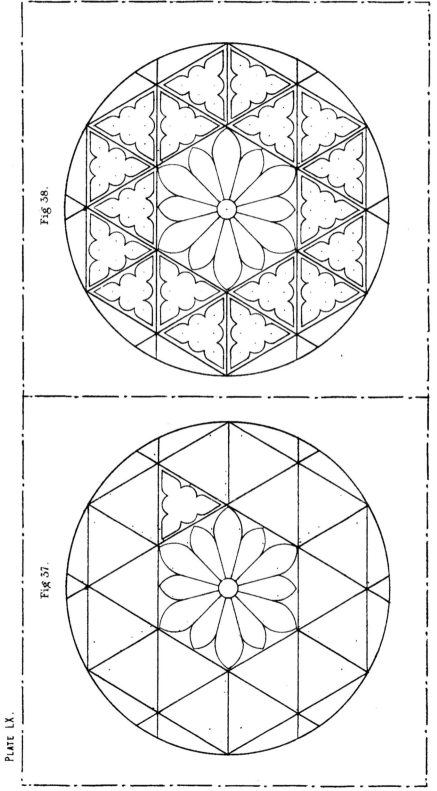

Fig 38.

Fig 37.

Plate LX.

Sc. 1.—Here we have a plain scale of ⅜ of an inch to the foot, in which a length equal to ⅜ of an inch, is marked off any number of times along a line, according to the space that is available, or according to the greatest measurement that it is required to take from the scale; the first portion is divided into 12 equal parts, which represent inches, numbered from point zero, which is placed at the right extremity of the first division to the left, while the divisions corresponding to feet, are numbered to the right. To take off any measurement, as 5' 2", from this scale, one point of the compasses must be placed on point 5 of the feet, and the other point extended to point 2 of the inches; and similarly, with regard to any other measurement that may be required.

Sc. 2.—Is a scale in which 7 inches represent a mile; it is divided, first, into 8 equal parts, to represent furlongs, and the first part is divided into 4, and one fourth into 10 parts, to represent poles. This is numbered, as in Fig. 1, to the left and right from o, and measurements in furlongs and poles may be taken from it, in the same way as feet and inches in Fig. 1.

Sc. 3.—Here it is required to make a scale of $\frac{1}{27}$ of actual size; the first step, which belongs rather to arithmetic than to geometry, is to find $\frac{1}{27}$ of some measurement that might be used in the length we have at our disposal, and that is at the same time sufficiently large for accurate measurement; thus, $\frac{1}{27}$ of a mile would be too much, and $\frac{1}{27}$ of a yard is smaller than is necessary, but $\frac{1}{27}$ of 3 yards, which is 4 inches, gives us a very available length. This 4 inches we mark off on the line, and divide it first into 3, to represent yards, and one of these yards into feet, and again one foot into inches. In this figure we have the scale extended to measure a total length of 5 yards, by adding the divisions of the four inches to that line produced. Any measurement may be taken from this scale, as from the previous ones. (Note 1.)

Note 1.—It is perhaps scarcely necessary to observe, that this is found by reducing 3 yards to inches, and dividing them by 27, which gives the 4 inches as a result, to arrive at the choice of 3 yards as a quantity to be divided, must be a matter of ex-

Sc. 4.—Is a scale of $\frac{1}{2500}$, showing poles and furlongs; in that, as in the previous scale, it was necessary to find the $\frac{1}{2500}$ of some available length, and we find that the $\frac{1}{2500}$ of a quarter of a mile, which is equal to 6.33 inches, is a suitable length; marking off, therefore, the actual length of 6.33 inches, we divide it, first, into 2 parts for furlongs, and one of these into 8 parts, to represent 5 poles each, and one of these parts into 5, to give us single poles. (Note 2.)

Sc. 5.—A scale of $\frac{1}{500000}$ is constructed similarly to Sc. 3 and 4; it being ascertained that $\frac{1}{500000}$ of 10 miles is equal to 1.2 of an inch, that measurement is marked on the line, and repeated 5 times to stand for a total of 50 miles, the first part is divided into 10, as single miles. (Note 3.)

Sc. 6.—Is constructed to represent tens of yards, and hundreds of yards, in the same proportion to actual size as 13 inches bears to 1,000 yards; as we could not measure 13 inches on our paper, and finding that 5.2 inches bears the same proportion to 400 yards that 13 inches bears to 1000 yards (Note 4), we measure off the 5.2 inches, divide it into 8 parts, to represent 50 yards each, and divide one of these parts into 5, to represent tens of yards. The scale is extended to measure 500 yards.

Sc. 7.—In this, similarly to the last, yards are marked in the proportion of 154 yards to 7.1 inches, by finding that 100 yards would bear the same proportion to 4.61 inches, the scale being extended to measure 140 yards.

Sc. 8.—Is a scale showing poles and furlongs in the same relation to actual size as 2.47 inches to 1000 yards (Note 5); it is constructed by finding that 4.35 inches is in the same proportion to one mile, and dividing the 4.35 inches accordingly.

periment or consideration, though it must be well understood, that the $\frac{1}{27}$ of another measurement would do quite as well, if it were an available length.

Note 2.—To find this we reduce a mile to inches, which number 63,360, divided by 2,500, gives us 25.32 inches; this would be too great a length for our scale, being

PLATE LXI.

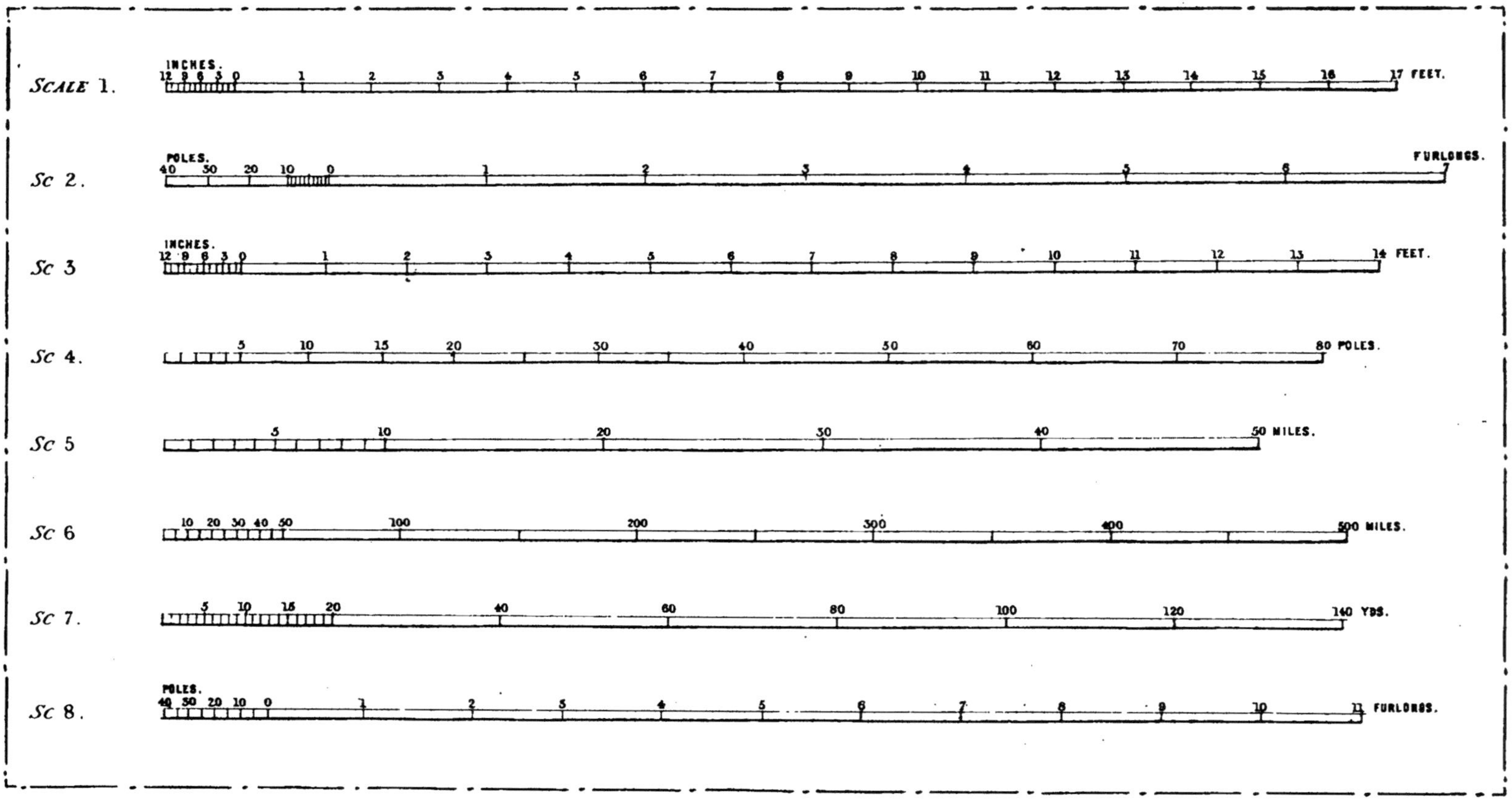

SCALE 1.

Sc 2.

Sc 3.

Sc 4.

Sc 5.

Sc 6.

Sc 7.

Sc 8.

PART III.—APPLICATIONS OF GEOMETRY.—PLATE LXII. (SCALES.)

Sc. 9.—When it is necessary to show very minute divisions, a diagonal scale may be used, similar to Fig. 3, Plate IIIa. We have here, a scale of 3 inches to a mile, divided into 8 equal parts, to show furlongs, and one part representing a furlong, subdividing into fortieths to show poles, by means of diagonal lines; it is used in the same way as Fig. 3, Plate IIIa.

Sc. 10.—Is a scale of $\frac{1}{5280}$ of actual size, or 2·25 inches to a furlong, divided to tens of yards, and single yards, by a diagonal line.

Sc. 11.—To make a scale of comparative French metres, to any given English scale, as the scale of 1 inch to a foot, we must bear in mind, that the French standard metre is to all degrees of practical accuracy, equal to 39·37 inches, and therefore, we have only to find such a measurement as shall bear the same ratio to 39·37 inches that the English scale bears to

actual size ; thus, as 1" is to a foot, i.e., $\frac{1}{12}$, so 3·28" is to 39·37, i.e., $\frac{1}{12}$; so that setting off

Sc. 11a.—3·28" on our scale, and dividing it into ten equal parts, we have a French scale of decimetres, and adding a division equal to a decimetre, and dividing it into ten equal parts, we show centimetres. The scale is produced to measure two metres.

Sc. 12.—Similarly, to make a scale of French hectometres, comparative to an English scale, as the given scale of 4¼ inches to a mile, in which furlongs and poles are shown. As 4·25" is $\frac{1}{14080}$ of a mile, so 2·64" is in the same proportion to a kilometre, which is equal to 39,370·79 inches, or 1,000 metres, the amount 2·64 being easily found by the sum—as 1 mile : 4·25" :: 1,000 metres, i.e., 39,370·79 inches : 2·64".

Sc. 12a.—The measurement of 2·64" is set off on the scale, to show kilometres, and divided and sub-divided to show hectometres and decametres.

Note 4.—This is found by a sum of simple proportion, as 1,000 yards : 400 yards :: 13 inches ; multiplying the 13 inches by 400, and dividing the result by 1,000, which gives us the 5·2" as an answer. This 400 is chosen as the 3 yards (Note 1), by experiment or consideration, which tells us that though 13 inches may be too large a measurement for our scale, that about $\frac{1}{3}$ of that would be suitable, and we chose an even number rather than a less even one, because it may afterwards be more easily divided on our scale.

Note 5.—In this scale, where it was required to represent furlongs and poles, while the proportion was given as so much to a number of yards, it became necessary to find the same proportion to a mile, or some larger division of a mile than a yard ; thus—as 1,000 yards : 1,760 yards, the number in a mile, so 2·47 is to 4·35" found by multiplying the 2·47 by 1,760 and dividing the result by 1,000.

about four times as much as we require, so we divide the 25·32' into four, which gives us the 6·33 inches to a quarter mile, or two furlongs.

This fraction, $\frac{1}{126720}$, is called the representative fraction of the scale; it is sometimes required to find the representative fraction of a given scale, where a certain actual measurement represents a stated length, as in Scs. 1 and 2; this may always be done by dividing any portion of the scale into whatever it represents, as $\frac{1}{2}$ of an inch into 1 foot (Sc. I), gives us $\frac{1}{24}$ as the representative fraction, and $\frac{1}{7}$" to one mile : Sc. 2 gives us by dividing 63,360, the number of inches in a mile by 7, a representative fraction of $\frac{1}{9051}$, the 3 inches remaining over, being set aside, as making no practical difference.

Note 3.—As 526,000 cannot be divided into 63,360 (the number of inches in a mile), we multiply that number by 10, which gives us the number of inches in 10 miles, and dividing this by 526,000, we have 1·2 inches as a result, which will consequently be the $\frac{1}{10000}$ of ten miles.

PLATE LXII.

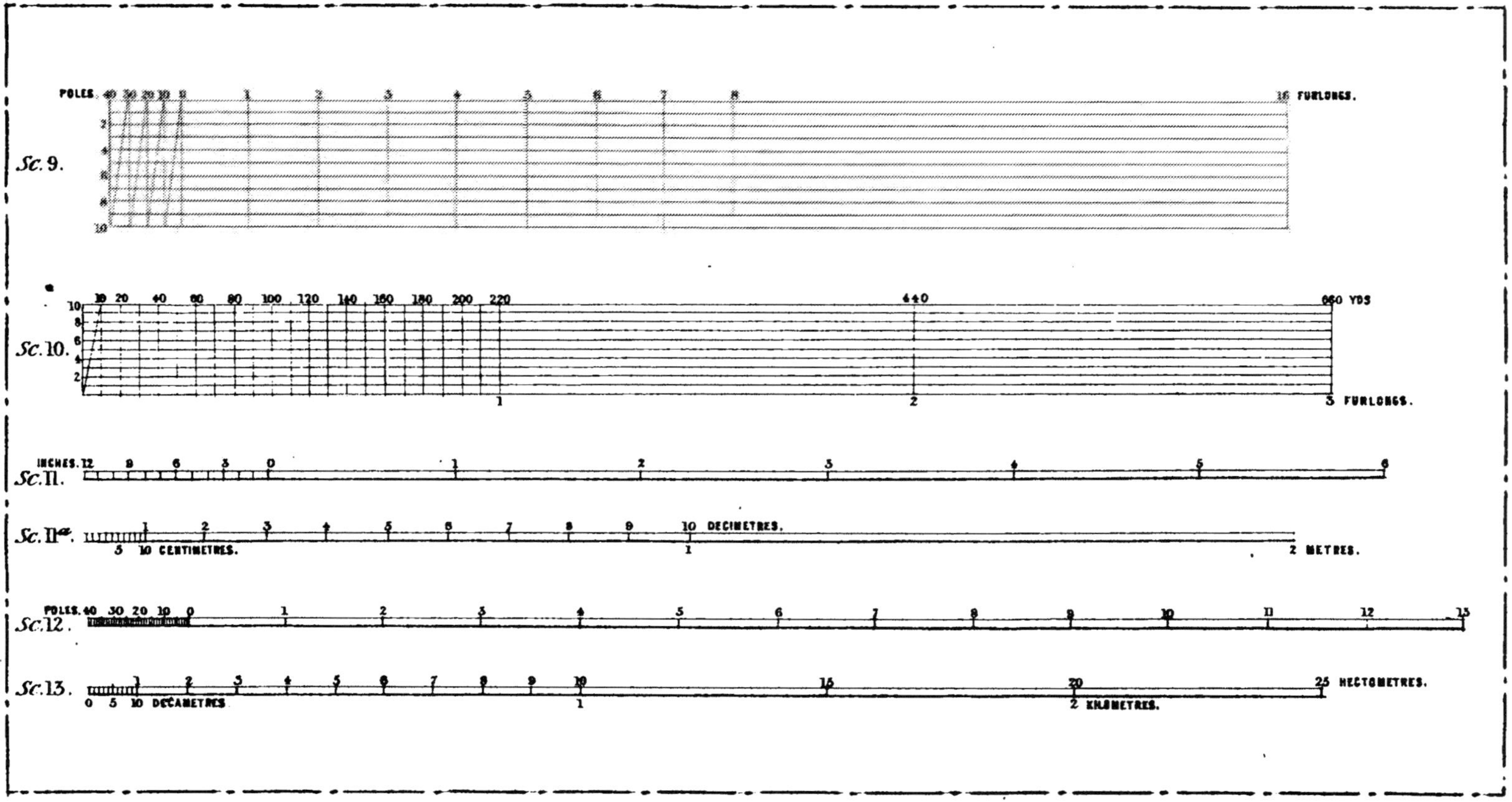
Sc. 9.
POLES. 40 30 20 10 0 1 2 3 4 5 6 7 8 16 FURLONGS.
2 4 6 8 10
Sc. 10.
10 20 40 60 80 100 120 140 160 180 200 220 440 660 YDS
10 8 6 4 2
1 2 3 FURLONGS.
Sc. 11.
INCHES. 12 9 6 3 0 1 2 3 4 5 6
Sc. 11a.
1 2 3 4 5 6 7 8 9 10 DECIMETRES.
5 10 CENTIMETRES. 1 2 METRES.
Sc. 12.
POLES. 40 30 20 10 0 1 2 3 4 5 6 7 8 9 10 11 12 13
Sc. 13.
1 2 3 4 5 6 7 8 9 10 15 20 25 HECTOMETRES.
0 5 10 DECAMETRES. 1 2 KILOMETRES.

PART III.—APPLICATIONS OF GEOMETRY.—PLATE LXIII. (SCALES.)

Sc. 13.—In constructing an English scale comparative to a given French scale, we have less difficulty, from the decimal system of numeration, at once showing the proportion that the scale bears to actual size; or, in other words, the representative fraction; thus, in this scale, we have one decimetre to a metre,

Sc. 13A.—in other words $\frac{1}{10}$ actual size; and if we mark 3·6 inches on our English scale, it will give us the comparative measurement for a yard; this should be divided into 3 equal parts for feet, and one foot divided into inches.

Sc. 14.—Here we have a French scale, of one decimetre to a hectometre, or one metre to a kilometre, or $\frac{1}{1000}$ actual size; from which, a comparative English scale is made, without using any English measurement, by these means—As 36·37 inches : 100

Sc. 14A.—centimetres :: 36 inches : 91·4 centimetres; mark off, therefore, on the English scale, a measurement of 91·4 centimetres, of the French scale, and it will stand for one hundred yards,. English, by which the scale is completed.

Sc. 15.—Is an English scale of 6" to a geographical degree, or $\frac{1}{101111}$ of real measurement, divided to show geographical miles.

The following scales are all comparative to the scale of 6" to a degree; thus, as there are 69·1 statute miles to a degree,

Sc. 15A.—we have in scale 15A,—as 69·1 miles : 6 inches : 50 miles to 4·33 inches; therefore, we set off a measurement of 4·33, and divide it, first into 5 parts, and then $\frac{1}{3}$ into 10, so completing the scale to show statute miles.

Sc. 15B.—There are 104·5 Russian versts to a degree, and to make a scale of versts, comparative to Sc. 15, we have—as 104·5 versts : 6" :: 100 versts : 5·74", and we mark off 5·74" on our scale, and divide it into 10 parts, and $\frac{1}{10}$ into 10.

Sc. 15C.—There are 75 Turkish berri to a degree, and therefore, at the ratio of 6" to a degree, we have 50 berri to 4", from which we set out our scale.

Sc. 15D.—There are 28·5 French post leagues to a degree, and therefore—as 28·5 French post leagues : 6" :: 20 leagues : 4·17", by which we set out the scale.

Note 1.—The student must observe, that upon the same principles, any number and varieties of scales may be made.

END OF PLANE GEOMETRY.

PLATE LXIII.

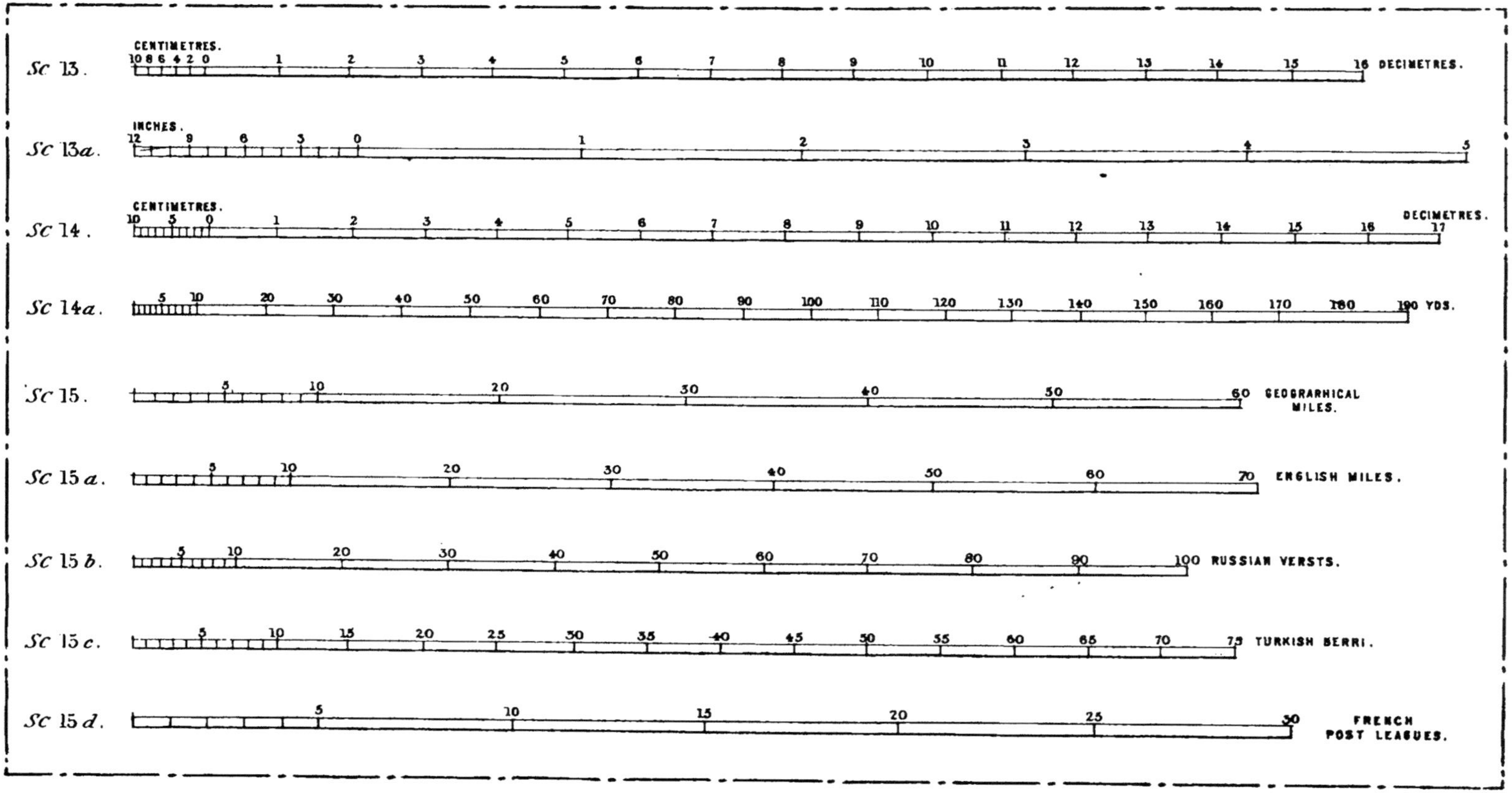

Sc 13.

CENTIMETRES.
10 8 6 4 2 0   1   2   3   4   5   6   7   8   9   10   11   12   13   14   15   16 DECIMETRES.

Sc 13a.

INCHES.
12   9   6   3   0   1   2   3   4   5 DECIMETRES.

Sc 14.

CENTIMETRES.
10   5   0   1   2   3   4   5   6   7   8   9   10   11   12   13   14   15   16   17 DECIMETRES.

Sc 14a.

10   20   30   40   50   60   70   80   90   100   110   120   130   140   150   160   170   180   190 YDS.

Sc 15.

5   10   20   30   40   50   60 GEOGRAPHICAL MILES.

Sc 15a.

5   10   20   30   40   50   60   70 ENGLISH MILES.

Sc 15b.

5   10   20   30   40   50   60   70   80   90   100 RUSSIAN VERSTS.

Sc 15c.

5   10   15   20   25   30   35   40   45   50   55   60   65   70   75 TURKISH BERRI.

Sc 15d.

5   10   15   20   25   30 FRENCH POST LEAGUES.

APPENDIX.

PART IV.—ELEMENTARY PROJECTION.

DEFINITIONS OF THE MOST COMMON GEOMETRIC SOLIDS.

A tetrahedron, is a solid bounded by four equal eqrtl. triangles. Four is the smallest number of plane sides, that a solid can have.

A pyramid, is a solid bounded by a number of planes, passing from, and forming equal angles to its base, and meeting in a point, which is called its apex or vertex ; the base may be any regular geometric figure, as a square, a hexagon, &c., from which, the pyramid is called a square, pyramid, or a hexagonal pyramid, &c.

A cube, is a solid bounded by six equal squares.

A parallelopiped, is a solid bounded by six plain rectangular surfaces, each opposite two, being equal and parl. to each other.

An octahedron, is a solid bounded by eight equal eqrtl. triangles, according to strict geometry, but it is frequently used to express a solid bounded by eight isosceles triangles.

An icosahedron, is a solid bounded by twenty equal eqrtl. triangles.

A dodecahedron, is a solid bounded by twelve equal regular pentagons.

A prism, is a solid bounded by two plane figures parl. and equal to each other, and a series of rectangular surfaces, joining the edges of these figures to each other.

A cylinder, is a solid such as would be generated by the motion of a rectangle, revolving round one of its own sides, from which it has two circular ends or bases, and one curved surface. A right line passing through the centres of both ends, is its axis, and the diameter of the base, is the diameter of the cylinder.*

A cone, is a solid such as would be generated by the motion of a right angled triangle, revolving round one of the sides of the right angle, from which it has one plane circular surface or base, and a curvilinear surface culminating in a point or apex, which is the extremity of the side of the triangle upon which it revolves ; such line, i.e., a line from the apex to the centre of the base, is the axis of the cone. Any right line from the edge of the base, to the apex of the cone, is spoken of as its side. A cone is termed acute, or obtuse, or right angled, according to the angle formed by two lines passing from opposite points on the edge of the base to the apex.

A sphere, is a solid of one unvarying curved surface, such as would be generated by the motion of a semicircle rotating upon its diameter. The centre of the diameter is the centre of the sphere, and any right line, passing through the centre, and bounded by the curved surface, is a diameter of the sphere ; the axis of the sphere is any diameter.

* Figures in all other respects similar to these, but having their bases not at right angles to their axes, are termed oblique, as an oblique cone, or an oblique prism, or cylinder.

Orthographic projection, is the art of representing upon a plane, or flat surface, figures which lie in planes not parallel to it, or solid objects, having the dimension of thickness, as well as length and breadth.

Solids are bounded, either by plane or curvilinear surfaces or both combined.

The delineation upon the plane is the projection, and the lines by which such delineation is found, are called projectors.

To give a full idea of a solid, it is necessary to have projections upon two planes, thus we have a projection upon a horizontal plane, to show the width and depth (from front to back), and a projection upon a vertical plane, to show the height or heights.

The plane upon which the projections are made, is called the plane of projection; those most constantly used, are the horizontal, and the vertical planes, these are considered together, and are termed co-ordinate planes.

In Fig. 1 these planes are represented by the aid of another science, called perspective, which appeals more directly to the uninitiated, since it represents objects as they would appear to the eye, but to indicate the position of the plane, it has been necessary to use lines, as if the plane were of a definite form; it must, however be understood that the plane is supposed to extend indefinitely in every direction.

The form A B C D, represents the vertical plane, and A D E F represents the horizontal plane; the line A D, in which both planes meet, or intersect, is the line of intersection, or intersecting line, which is its proper title, although it is by some, somewhat arbitrarily, called the line X Y, and sometimes called the ground line.

In the same figure, we have represented a point in space (P), and a solid figure (K), together with their projections, upon both planes, and the projectors.

A line or projector, being dropped vertically from point P, comes in contact with the horizontal plane, in point G, which is its projection upon

that plane, and a line from P, drawn perpendicular to the vertical plane, meets it in H, giving another projection of the same point.

All projections on the horizontal plane, are called plans, and all projections on the vertical plane, are called elevations.

Shape I, is the elevation of the solid K, and L is its plan; this solid is supposed to be placed, with four of its surfaces or planes, perpendicular to each plane of projection, thus by producing the edges of those planes we have projectors, which give us four lines in each projection, which lines each represent a plane of the solid; the four lines of the plan representing the four vertical surfaces, and the lines of the elevation representing two of its vertical, and two horizontal surfaces.

In more advanced studies of this science, the planes are used as passing through each other, thus forming four dihedral angles, either one or all of which may be used at the same time, but in the elementary study, only the upper and nearer angle is used, as in the illustrations, the object being supposed to be always in front of the vertical plane, and above the horizontal plane.

Fig. 2 shows two solids with their plans and elevations and projections, as in Fig. 1, but in addition, the horizontal plane is shown as turned down, until it coincides with the vertical plane. This is to illustrate how it is used upon a sheet of paper, when both plans have to be represented upon one surface; the horizontal plane A B C D, must be realised as turned down upon line A B, carrying with it the plans C and C′ which are shown also in shapes E and E′.

It will be seen, that the points in the plans E and E′ are carried back to the vertical plane, by lines perpendicular to it, cutting the intersecting line in points O O O, &c., and that vertical lines through these points contain the points in the elevations; furthermore, it must be observed that these lines which are perpendicular to the intersecting line become, of a necessity, continuous with the lines on the vertical plane, when the horizontal plane is turned down to coincide with the vertical plane, and from this we get the very important axiom, that the elevation of every point, is vertically above its plan, and *vice versa*, the plan of every point, is vertically below its elevation.

PLATE LXIV.

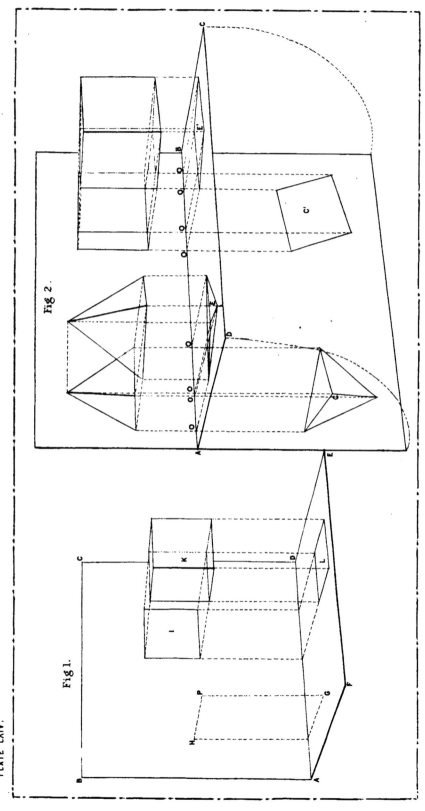

Fig 1.

Fig 2.

In this, as in all the succeeding plates, the line l L represents the intersection of the two planes, all above it being the vertical plane, and the portion of the paper below it, the hortl. plane. The student should observe, that the intersecting line at the same time, stands for both planes; for while it is the elevation of the hortl. plane, it is also the plan of the vertl. plane.

Fig. 1.—Is the plan and elevation of a cube, having one of its faces in a hortl. plane, and one face parl. to the vertl. plane, by drawing lines up from its sides perpr. to l L, and above it, and marking them off in line A B, equal one side of the plan, and drawing the hortls. A C and B D, we have the elevation of the same cube; the height A B, may be set up from the l L or not, as it may be required, to show its position relative to the hortl. plan, in the same way, the plan may be placed at any convenient distance from l L, as it gives the position relative to the vertical plane.

Fig. 2.—Is a second plan and elevation of the same cube, which, similarly to the last, is standing on one face on a hortl. plane, but in this, the cube is placed at an angle with the vertical plane, i.e., its faces if produced as planes, would make dihedral angles with the vertl. plane. In this case, the projectors are drawn up perpr. to the l L, from the four angles of the plan, and then cut off in E F, at a height from the l L, equal to one side of the plan, and the line F G drawn to complete the elevation. The edge H is shown as a dotted line, or omitted because it is really invisible, since the solid is in front of it.

Fig. 3.—Is the plan and elevation of a parallelopiped or rectangular slab, it has one face parl. to the hortl. plane and its vertl. faces, making certain angles with the vertl. plane, these sides, as in the case of the last cube, may be any angles required, the side to make the given angle being drawn first, and the rectangle constructed upon it as before. All the corners or angles of the plan, are carried up by lines perpr. to the l L, and the given thickness marked off, as may be required in points, and K M at any suitable distance from the l L; as its lower and upper surfaces are both parl. to the hortl. plane, they will be represented in the elevation, by hortl. lines through points K and M, to cut all the lines drawn up from the points in the plan. The back edge N, is represented by a dotted line.

Fig. 4.—Is a plan and elevation of a hexagonal prism standing upon one end, with two of its vertl. faces perpr. to the vertl. plane, from which, when we make the elevation by drawing the vertls. up and cutting them off at any required height, as the length of the prism, we find the elevation shows only two sides of the prism.

Fig. 5.—Is a second plan and elevation of the same prism. In this case it is placed with one face making an angle of 45° to the vertl. plane, by which, when we complete the elevation, we have three sides visible, and the two invisible edges represented by the dotted lines.

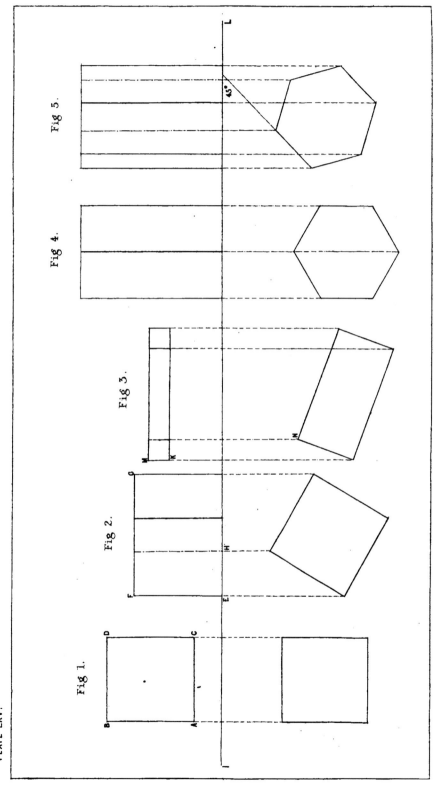

PLATE LXV.

Fig 1.

Fig 2.

Fig 3.

Fig 4.

Fig 5.

In making the projections of solids, it is necessary to make the plan or elevation first, according to the positions in which they may be placed; since it is always necessary to make that projection first, which gives some part, or parts of the figure as they really are; but, in some cases, both projections are so real in the dimensions, that it is immaterial which is commenced first; as, for instance, the first figure in the last plate, where the cube is, in both elevation and plan, represented by a square; and in the first figure of this plate, where we have a cylinder standing on its base, in which case, either the

FIG. 1.—rectangle showing its elevation might be completed first, and the diameter of the circle found, by dropping the sides down, and drawing a hortl. line, wherever it may be convenient, across them, for the diameter; or the plan might be constructed first, and the elevation made from it. The cone, Fig. 3, is an

Fig. 3.—exactly similar case, the plan may be made from the elevation, or *vice versâ.*

FIG. 2.—Is the plan and elevation of a square pyramid, standing on its base; it is in this case necessary to begin with the plan, because it is placed with one edge of the base, as A B, making an angle, in this case of 52°; in drawing the plan of a pyramid in this position, the centre of the base is found, as C, and lines drawn from the angles to the centre, to show the edges of the inclined sides; this point, C, is carried up by a

perpr., to give the apex, C^1, which may be marked at any height that is desired, and the same may be said of the cone, for it must be remembered, that the lengths of the sides in cones, and pyramids, or cylinders, bear no fixed ratio to the base.

FIG. 4.—In the case of a pentagonal pyramid, standing on its base, it is absolutely necessary to begin with the plan, because we cannot otherwise find the width of the sides; in this example, the base is placed with one of its edges A B, making an angle of 20°, to the I L, which is taken as its given position, and the pentagon constructed upon A B,—its centre, C, is then found, and the edges drawn from the angles. The elevation is completed as in the previous figures.

FIG. 5.—This is a projection of a square frame, composed of square bars; it is placed, with what would be termed one face of the frame, inclined at an angle of 60° to the vertical plane, one side of the frame resting on the horizontal plane. When the plan is drawn in the given position, and of the required size, the thickness of the upright sides is indicated by the dotted lines in the plan; and for the elevation, all the points in the plan are drawn up, and it being a square frame, the height is the same as the length, A B, of the plan, and all the bars being square, their heights are the same as A C

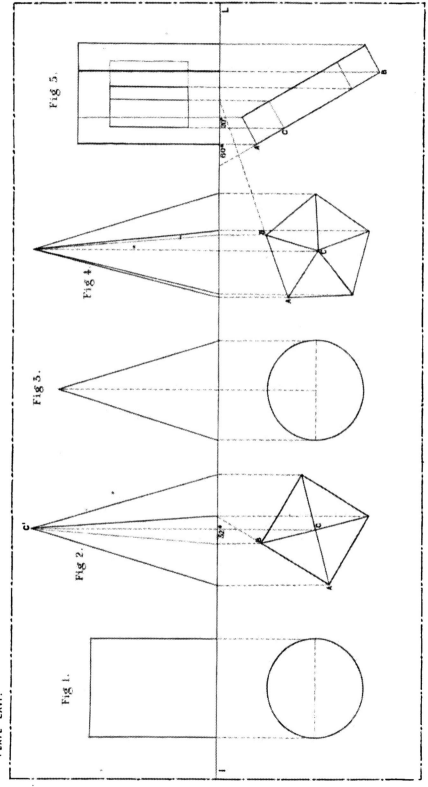

PLATE LXVI.

Fig 1.

Fig 2.

Fig 3.

Fig 4.

Fig 5.

FIG. 1.—In some cases, it is possible for the student to remember that it is possible for the properties of a solid to enable him to make a plan or elevation, to a certain extent by induction; as, for example, in Fig. 1 of this Plate, we have the plan and elevation, of a triangular prism, resting on one of its sides; although strictly speaking, the plan would be made from the elevation, yet, if the student remembers, that since the sides are equal, the ridge, or one edge, A, must be, in projection, equidistant from the edges B and C, he will see that he can make the plan, by drawing the lines B and C, at the given width of the side of the prism, and drawing line A, bisecting the distance between them; the ends of prisms being taken as at right angles to the sides.

FIG. 2.—In making the projection of an octagonal prism resting on one side, it is necessary to begin with an elevation, since the widths of the outer sides can only so be found; thus, in Fig. 2, the octagon is first constructed upon the line A B, as the given width of the side, and lines drawn down perpr. to the I L, and cut off at the required length in the plan.

FIG. 3.—Shows the plan and elevation of a truncated pyramid; the widths of the base and top, and its height being given; the elevation is first drawn, according to those data, remembering that the centre of the top must be over the centre of the base; in making the plan, when the plan of the base, A B C D, is found, the lines dropped from the top will cut the diagonals of the square in the points E F G H, and it will be completed by joining the points E F and G H.

FIG. 4.—This is a V shoot, or trough, placed so that it rests upon its lowest edge; in this case the elevation is made first, and the plan is constructed from it.

FIG. 5.—Shows what may be termed a compound solid; it is composed of a short cylinder, with two cones, placed with the centres of their bases coincident with the centres of the ends of the cylinder; thus, they have a common axis, A B; when placed in this position, with the axis perpr. to the vertical plane, either the plan or the elevation may be drawn first; the dimensions are, of course, given, since they may be varied in all these figures, at pleasure.

Plate LXVII.

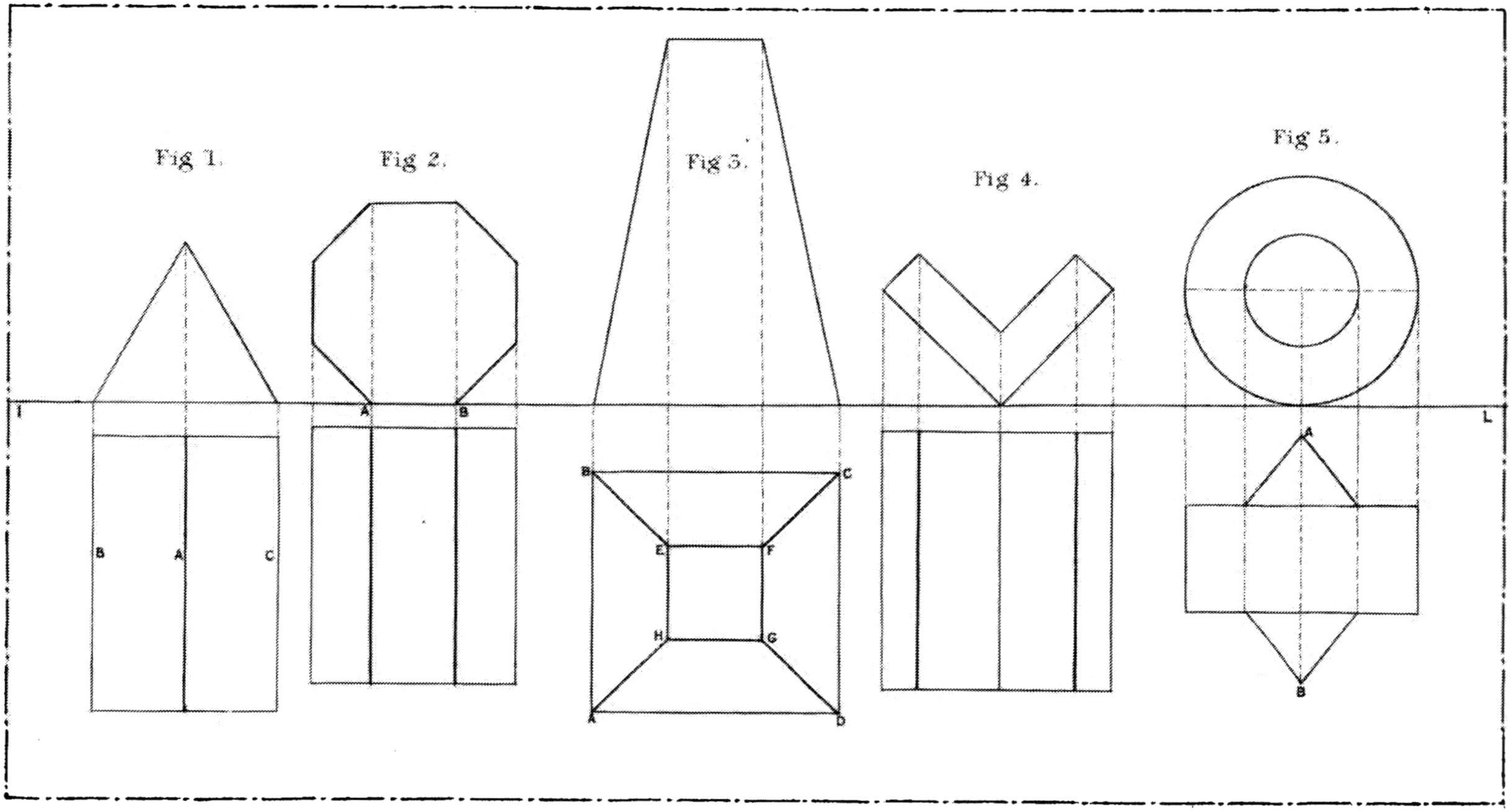
Fig 1.
Fig 2.
Fig 3.
Fig 4.
Fig 5.
I
L
A
B
B
A
C
A
B
C
E
F
H
G
D
A
B

In all the foregoing examples the solids have been placed in the simplest possible positions, as illustrations of plans and elevations: it will be evident, that each solid might be placed in an endless variety of positions, having various projections; some of the easier of these variations of positions, with their projections, are exemplified in this plate.

Fig. 1.—Here a short cylinder is placed on its side, with the planes of the ends making angles of 45° to the vertical plane; we are obliged to make the plan first, in the required position, and then to project up the two lines, A B and C D, for the axes of the ellipse of the elevation;—line Z Y is bisected in E, and point E projected up, being made in B F and F A, equal Y E and E Z; a horizontal line through F, cut by the projectors from Z and Y, gives the minor axis. The ellipse may be completed by either of the methods given in the Plane Geometry, or, by the method of ordinates here indicated, and the half of the second ellipse found in a similar way.

Fig. 2.—The projections requisite to obtain the plan and elevation of a pyramid, will vary according to the data from which they have to be constructed, and the position required; here the side and form of the base, and the height, are given, and a plan and elevation required; when the pyramid is lying on one side; construct the plan upon a line, A B, perpr. to the I L, and from it draw the elevation. From point C, rotate points D, E, and F, as indicated by the arcs, making G H equal D E, and joining points H and C; we then have the elevation of the pyramid, turned upon its side; to obtain the plan from this, drop lines from all the points in the elevation, and cut them, by lines from the corresponding points in the first plan.

Fig. 3.—Gives a cone, treated exactly the same as the last figure, except that the point A is chosen at a distance from the first elevation; A B is made equal Z Y, and from A, with radius Z X, and from B, with radius Z Y, arcs are described cutting in E,—the ellipse of the plan is found similarly to Fig. 1.

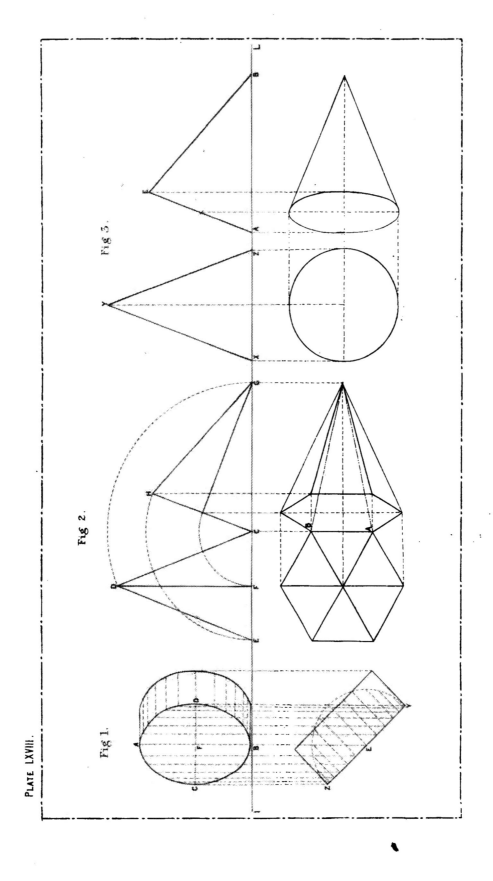

PLATE LXVIII.

Fig 1.

Fig 2.

Fig 3.

Among the varying data for the positions of solids, which may be required to be shown in projection, perhaps the most common, are when one edge of a face is given at a certain angle to the vertl., and at the same time, the inclination to the hortl. of a face, of which the given edge is a side.

FIG. 1.—In this example, the edge of the base A B, and the height C D, of a pentagonal pyramid is given; which it is required to project, when one edge of the base rests on the ground, making an angle of 45° to the vertl., and the base is inclined at 45° to the hortl. On A B construct the pentagon A B E F G,—lines drawn from the angles to the centre H show the ridges of the pyramid,—draw a line K L at 45° to the I L equal to I F,—on K L construct a side elevation of the pyramid.

FIG. 1 a.—Line a, b being drawn at 45° to the I L, and being bisected by a perpr.; points 1, 2, and 3 in Fig. 1 give us the distances Z, Y, and X, and line G E gives us the width. From each point in this plan a perpr. is carried up, and cut by a hortl. from each corresponding point in the side elevation, by which means we obtain the elevation required in Fig. 1a. This method may be adopted similarly, in most solids where the data are similar.

FIG. 2.—To give the projections of a cube, when one face is inclined at 45° to the hortl. plane, and one edge upon which it rests on the hortl. plane is inclined at 30° to the vertl. plane. Line A B is drawn at 30° to the I L, and to save space and

time, the side elevation of the cube, C D E F is drawn in the required position to Z Y, as representing the hortl. plane,— Z Y must be at right angles to A B,—and by this the plan is first constructed; from each point in the plan a perpr. is carried up, and the heights measured from the I L, equal to the distances of the points from Z Y.

FIG. 3.—To give the projections of an octahedron resting on one of its sides, when its axis is in a vertl. plane, making an angle of 45° to the vertl. plane.

It must be remembered, that a regular octahedron, is composed of two square pyramids joined at their bases, and having eqtrl. triangles as faces.

To construct the side elevation, which is in this case drawn separately to avoid confusion, although it might otherwise be drawn as in the cube; A B C being taken as one face of the octahedron, draw A D perpr. to B C,—from any point as Z with A D as radius describe an arc to cut the I L in Y,—from Y with radius A B cut the first arc in X,—from Y and X with radius Y Z describe arcs cutting in W.

FIG. 3A.—Having the side elevation, the plan of Fig. 3a is constructed from it in the same way as in Fig. 1a, the line V U being drawn at 45° to the I L, and the elevation is found, by drawing up the perprs. from each point in the plan, and cutting them off by hortls., from the corresponding points in the side elevation, as in Fig. 1.

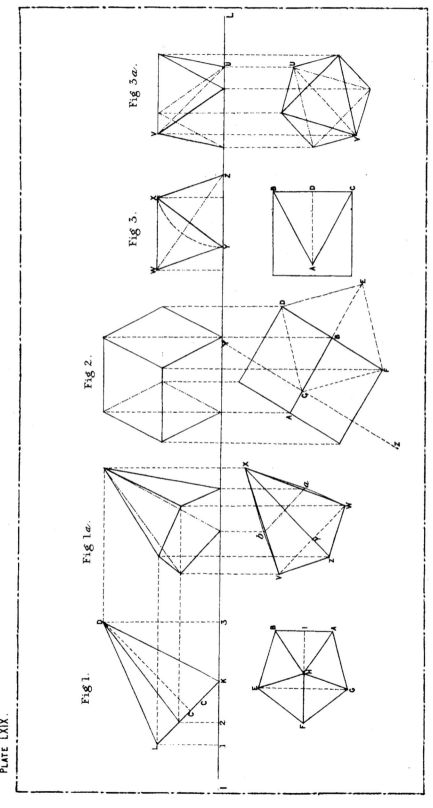

PLATE LXIX.

Fig 1.

Fig 1a.

Fig 2.

Fig 3.

Fig 3a.

Besides the numberless cases, of the projections of whole solids, it is frequently necessary, to show what is called a section, or a cutting through. In this case, a line called a section line, or line of section, is marked across the solid in any required position, and the second projection is made exactly as if all that portion of the solid, to the left of the section line, were removed; or, in a hortl. section, all that portion of the solid above the section line.

Fig. 1.—In this, the elevation of a cube, lines S¹ and S² are section lines; and Figs. 2 and 3 show these sections worked out.

Fig. 2.—To obtain this projection, as the section is vertical, it is represented in the plan, by line a b, which cuts across the plan in those points,—a line D L, called a datum line, is then drawn across the plan, at any convenient point, parallel to the I L, and another datum line d l, is drawn on the vertl. plane,—these lines give us the positions of every point in the solid, as so much to the right or left, remembering that the section is always supposed to be viewed, in the direction indicated by the arrow heads;—having then the line a b, cutting the D L in point ɢ, we measure to the left and right of line d l, the distances c a, and c b, in points e and f, and to show the parts of the other solid, we measure g h to the left of d l, in i, and k m to the right of d l, in o.

Fig. 3.—This figure differs from the last, inasmuch as we have to be careful of the hortl. level, of each point in the section;

thus points 1, 2, 3, and 4, where the section cuts each edge of the solid, are drawn horizontally across the d l, and the same points being found in the plane their positions to the right or left of the D L, is marked in points 5, 5ᵃ, 6, 7, and 8, the vertical lines from points 6, 7, and 8, giving those parts of the solid not in section.

In making the plans from the sections, we have only a projection of the part of the solid, remembering always, that the distances from front to back, i.e., from Z toward Y, may be found in the original plan, in the distances from d towards l.

The sectional face is always indicated by a series of parallel lines, at any convenient angle, not to correspond too much with the outlines of the shapes.

Figs. 4 and 5.—Similarly, in this, line Sⁿ indicates a section, to be cut through a hexagonal pyramid;—the points where the section crosses the edges of the pyramid, as 1, 2, 3, 4, 5 and 6, are dropped upon the plan, in points 7, 8, 9, 10, 11, and 12;—the datum lines are then drawn, as before, and each point being drawn horizontally across, the distances to the right and left of the D L in the plan, are measured off at their respective levels, to the right and left of the d l, in the vertical plane; point 6 of the section, showing two points 12 and 12' in the plan, one on either side of the d l.

The plan, Fig. 5, is found as before, by dropping lines from the points in the elevation, to give the widths, and by measuring the distances from front to back, equal to the distances on the original plan from, D to L

PLATE LXX.

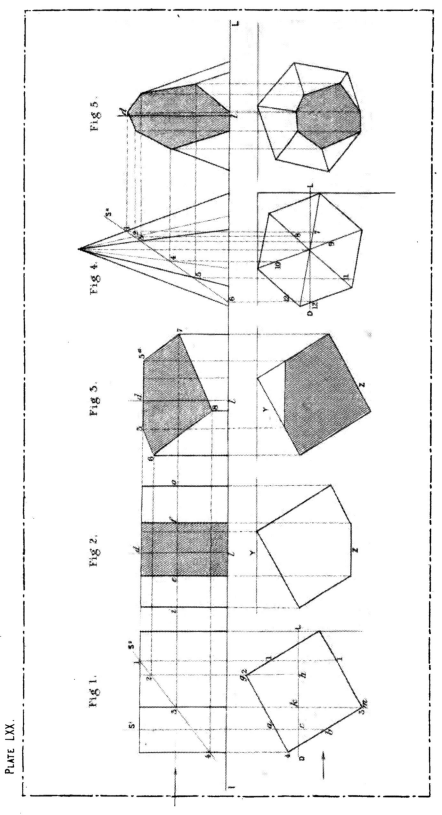

Fig 1.

Fig 2.

Fig 3.

Fig 4.

Fig 5.

desired point of the section, by merely dropping a line across the plan, unless the point is on the base of the cone. To overcome this difficulty, we may mark any series of points upon our section line, as points 1, 2, 3, &c., on line S A,—through each point as 1, draw a line parl. to the base of the cone as B C, B C being the diameter of the base of the cone B X C, when we describe a semicircle upon B C, and draw a line from point 1 perpl. to B C, to cut the semicircle in point D, we have the distance D I, as half the distance through the cone at point 1, and proceeding similarly, with all the other points, 2, 3, 4, &c., we shall have a series of widths which may be set off on either side of the D L of Fig. 6, at their respective levels, and through all these points so found, the curve must be drawn by hand. The extreme width, of that part of the cone not in section, may be found by the plan of Fig. 5,

Section 2, being an ellipse, is here shown as projected upon the hortl. plane, the axes are found by dropping lines from points E and F, and from G the bisection of E F, and finding the distance through the cone at point G as in point 1, this gives us the diameter H I, and the ellipse may be completed.

Section 1, is shown projected upon the hortl. plane, in the plan of Fig. 6, which is found in a similar manner to all other plans of section.

The sections of cones parallel to the axes of the cones, are hyperbolic curves, the sections parallel to the sides of the cones, are parabolic curves, and sections cutting through both sides of the cones, are elliptic, unless when parl. to the base, in which case they are circular.

FIG. 1.—Here, a section is cut through a cylinder and square block, in combination, the cylinder passing through the centre of the block, the axis of the cylinder being perpr. to the faces of the block; the section is found as in the preceding plate, a series of points as o, 1, 2, 3, &c., being chosen in the elevation, and lines dropped from these points, across the plan, by which means we obtain the measurements to the right and left of the datum line, for Fig. 2.

FIG. 2.—In showing a section of a compound solid, or two or more solids combined, care must be taken that the section lines of the different portions contrast with each other in direction, as in this example.

FIG. 3.—In some cases it is necessary to show the actual form, and dimensions of a sectional surface. To do so, draw a line, as A B,—upon A B set off all the actual distances as from o to 1, 1 to 2, &c., in points 6, 6, 6, &c.,—through these points draw lines perpr. to A B, and on each line, measure the same widths as in Fig. 2.

FIG. 4.—The section of a sphere is always a circle, and if the sphere is hollow, and the section cuts into the hollow, it always gives two circles, as the boundaries of the cut surface; these circles, however, would become ellipses, if we have a projection of the section, upon any plane that was not parl. to it, in which case, it is simplest to project the two axes of the ellipse, or in other words, the two diameters of the circles which form the axes of ellipse, and then to complete the ellipse as a plain figure.

FIG. 5.—In conic sections we find a difference from other figures, inasmuch as we cannot necessarily find the width, at any

PLATE LXXI.

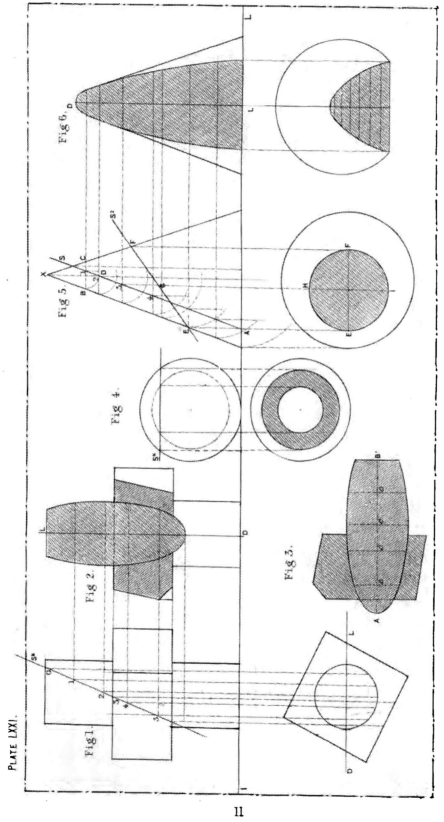

Fig 1.

Fig 2.

Fig 3.

Fig 4.

Fig 5.

Fig 6.

CPSIA information can be obtained at www.ICGtesting.com
Printed in the USA
LVOW052340261011

252217LV00011B/158/A

9 781437 063448